JACARANDA

GEOGRAPHY ALIVE **7**

VICTORIAN CURRICULUM | SECOND EDITION

JACARANDA
GEOGRAPHY ALIVE 7

VICTORIAN CURRICULUM | SECOND EDITION

JUDY MRAZ

CATHY BEDSON

BENJAMIN ROOD

CONTRIBUTING AUTHORS

ALEX SCOTT | ANNE DEMPSTER | JEANA KRIEWALDT
KATHRYN GIBSON | ALEX ROSSIMEL

jacaranda
A Wiley Brand

Second edition published 2020 by
John Wiley & Sons Australia, Ltd
42 McDougall Street, Milton, Qld 4064

First edition published 2017

Typeset in 11/14 pt TimesLTStd

ISBN: 978-0-7303-7990-4

Front cover image: © petekarici/Getty Images Australia

Illustrated by various artists, diacriTech and Wiley Composition Services

Typeset in India by diacriTech

Printed in Singapore by
Markono Print Media Pte Ltd

A catalogue record for this
book is available from the
National Library of Australia

10 9 8 7 6 5 4 3 2 1

CONTENTS

Geographical inquiry: What is my place like?

online only

HOW TO USE

the *Jacaranda Humanities Alive* resource suite

The ever-popular *Jacaranda Humanities Alive 7* has been re-published for the Victorian Curriculum. It is available as a single 4-in-1 title and as subject-specific titles: *Jacaranda History Alive 7*, *Jacaranda Geography Alive 7*, *Jacaranda Civics and Citizenship Alive 7* and *Jacaranda Economics and Business Alive 7*. The series is available across a number of digital formats: learnON, eBookPLUS, eGuidePLUS, PDF and iPad app.

Skills development is integrated throughout, and explicitly targeted through SkillBuilders and dedicated skills topics for History and Geography.

This suite of resources is designed to allow for differentiation, flexible teaching and multiple entry and exit points so teachers can *teach their class their way*.

Features

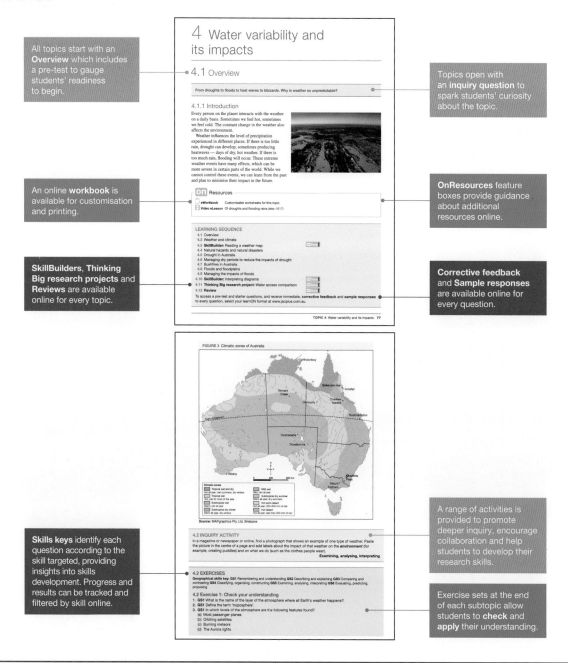

All topics start with an **Overview** which includes a pre-test to gauge students' readiness to begin.

An online **workbook** is available for customisation and printing.

SkillBuilders, Thinking Big research projects and **Reviews** are available online for every topic.

Topics open with an **inquiry question** to spark students' curiosity about the topic.

OnResources feature boxes provide guidance about additional resources online.

Corrective feedback and **Sample responses** are available online for every question.

Skills keys identify each question according to the skill targeted, providing insights into skills development. Progress and results can be tracked and filtered by skill online.

A range of activities is provided to promote deeper inquiry, encourage collaboration and help students to develop their research skills.

Exercise sets at the end of each subtopic allow students to **check** and **apply** their understanding.

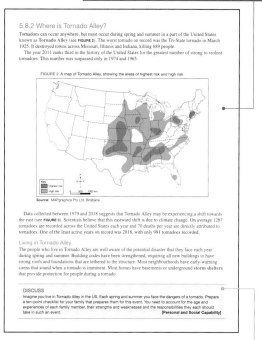

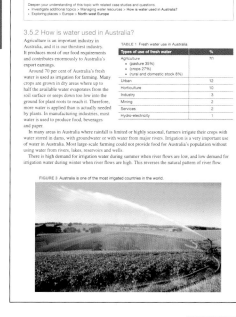

learn on

Jacaranda Humanities Alive learnON is an immersive digital learning platform that enables student and teacher connections, and tracks, monitors and reports progress for immediate insights into student learning and understanding.

It includes:

- a wide variety of embedded videos and interactivities
- questions that can be answered online, with sample responses and immediate, corrective feedback
- additional resources such as activities, an eWorkbook, worksheets, and more
- Thinking Big research projects
- SkillBuilders
- teachON, providing teachers with practical teaching advice, teacher-led videos and lesson plans.

teach on

Conveniently situated within the learnON format, teachON includes practical teaching advice, teacher-led videos and lesson plans, designed to support, save time and provide inspiration for teachers.

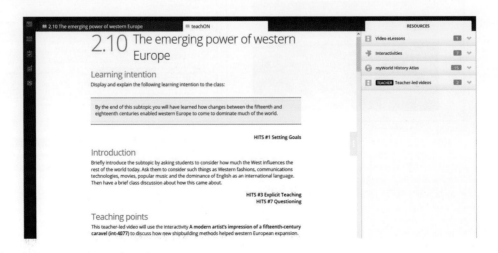

ACKNOWLEDGEMENTS

The authors and publisher would like to thank the following copyright holders, organisations and individuals for their assistance and for permission to reproduce copyright material in this book.

The Victorian Curriculum F–10 content elements are © VCAA, reproduced by permission. The VCAA does not endorse or make any warranties regarding this resource. The Victorian Curriculum F–10 and related content can be accessed directly at the VCAA website. VCAA does not endorse nor verify the accuracy of the information provided and accepts no responsibility for incomplete or inaccurate information. You can find the most up to date version of the Victorian Curriculum at http://victoriancurriculum.vcaa.vic.edu.au.

Images

• AAP Images: **88**/Jenny Evans; **100** (left, right)/Dave Acree; **72**/Graeme McCrabb; **125** (left, right)/Karin Calvert; **122** (left, right)/New Zealand Defence Force; **131**/Dave Hunt; **140** (top)/; **140** (bottom)/Max Becherer; **238**/AAP Newswire • Alamy Australia Pty Ltd: **60**/Jake Lyell; **125** (top right)/Bruce Miller; **241**/Danita Delimont; **14** (bottom right)/David Wall; **203** (right)/David Wall; **210** (bottom)/Liba Taylor; **55** (bottom)/Neil McAllister; **139**/Raynor Garey; **192**/Burleigh, Queensland; **27**/ARCTIC IMAGES; **27**/ARCTIC IMAGES; **224**/Bill Bachmann; **224**/Ariadne Van Zandbergen; **23** (bottom left)/Aurora Photos; **232** (e)/Australia; **226** (left)/Blend Images; **33** (top)/David Wall; **169** (bottom right)/David Wall; **232** (b)/Doug Houghton; **156** (top left)/Eric Nathan; **224**/Folio Images; **244** (middle right)/Friedrich Stark; **250** (a)/Geoff Smith; **3** (bottom left)/geogphotos; **244** (middle left)/guatebrian; **156** (right)/Guillem Lopez; **14**/Hemis; **155** (bottom right)/Ian Dagnall; **187**/Ian Nellist; **207**/imageBROKER; **244** (top)/imageBROKER; **244** (top right)/Irene Abdou; **244** (bottom left)/Johnny Greig Int; **232** (d)/Lukas Watschinger; **238** (bottom)/Marc Anderson; **146**/Michael Dwyer; **130** (bottom right)/model10; **144** (bottom)/NG Images; **29** (bottom)/Outback Australia; **57**/Peter Abell; **244** (bottom right)/Piti Anchalee; **174** (top)/Pulsar Imagens; **232** (f)/Robert Wyatt; **17** (bottom)/robertharding; **8**/robertharding; **250** (b)/Thomas Cockrem; **10**/ton koene; **242**/ZUMA Press, Inc.; **196**/AA World Travel Library; **226**/(A) Agencja Fotograficzna Caro • Angela Edmonds: **3** (top) • AQUASTAT: **48** (bottom)/Based on data from FAO. 2016 • Australian Bureau of Meteorology: **71**, **124**, **128** (bottom left), **128** (bottom right), **129** (top), **169** (top), **132**, **198**, **230**; **40** (top)/Andrew Treloar • Creative Commons: **25**, **54**, **9** (left), **143**, **148**, **149** (top right, top left), **159**, **189**, **204**, **203** (left); **28**/Mattinbgn; **20** (top left, bottom right)/Commonwealth of Australia; **163**/© Commonwealth of Australia National Archives of Australia 2015; **176**, **177**/© State of Victoria Department of Environment, Land, Water and Planning 2018; **138**/Vittorio A. Gensini & Harold E. Brooks; **45**/Adapted from Bureau of Meteorology and S Fatichi • Cynthia Wardle: **214** (bottom right and left) • Department of Foreign Affairs and Trade: **123**, **246**, **247** (a) • ECCA Nepal: **248** (a) • European Commission: **121**/2014 European Union • Getty Images: **154**/btrenkel; **150**/dlewis33; **77**/Julie Fletcher; **41** (bottom left)/M Swiet Production; **187**, **193**/Russell Tate; **241**/AFP; **134** (right)/AFP Australia; **58**/Bukbisj Candra Ismeth Bey Barcroft Media; **42** (bottom)/Dan Kamminga; **201** (bottom)/David Wall Photo; **40** (bottom)/DEA PUBBLI AER FOTO; **122**/Gary Williams; **210** (top)/George Holton; **120**/Handout; **247** (b)/Kaveh Kazemi; **181** (top)/Martin Cohen Wild About Australia; **106**/MUNIR UZ ZAMAN; **49**/Peter Walton Photography; **33** (bottom)/TED MEAD; **102** (bottom)/VANDERLEI ALMEIDA; **134** (left)/Visual China Group Australia; • Gold Coast SUNS: **192** (bottom)/Metricon Stadium • Grant Gibbs: **248** (b) • International Energy Agency: **21** (top) • iStockphoto: **4** (bottom right)/golubovy; **155** (top left)/Adam Kazmierski; **156** (bottom left)/Victor Maffe • John Wiley & Sons Australia, Ltd: **235**, (left) **160** (top), **48**/Data source: Australia Bureau of Statistics; **241** (bottom), **253** (top, bottom)/Based on data from Roland Berger and Statista • Johnny Haglund: **41** (bottom right)/Johnny Haglund • Karen Bowden: **3** (bottom right) • MAPgraphics: **81**, **83**, **137**, **164**, **7** (top), **17** (top), **44** (bottom), **189** (top), **81** • Margaret River & Districts Historical Society Inc: **202** (right) • NASA Earth Observatory: **144** (top right) • NASA: **145** (top); **134**/NASA Worldview; **100** (top left, top right)/NASA/GSFC/METI/ERSDAC/JAROS, and U.S./Japan ASTER Science Team • National Archives of Australia: **581** (top, bottom); **541** (top)/Terrance McGann • Newspix: **101**, Newspix; **104**/David Kapernick; **20** (bottom left)/Mark Calleja; **23** (bottom right)/Chris Hyde; **166** (bottom)/Bob Barker; **224**/Phil Hillyard • Nile Basin Initiative: **69** • Niranjan Casinader: **159**, **173**, **174** (bottom left and right), **175**, **212**, **213**, **214** (top right and left), **215**, **216**, **217**, **218**, **220** • Our World in Data: **244** (top)/Max Roser and Hannah Ritchie; **61** (top, bottom)/World Bank • PhotoDisc: **23** (top left); **23** (middle right)/Glen Allison • Picture Media: **40** (middle) • Queensland Bulk Water Supply Authority: **90** • Shutterstock: **220**/Catalin Lazar; **15**, **252** (top); **11**/Doidam 10; **16**/18011153; **35**/Taras Vyshnya; **37**/Johan Swanepoel; **51**/dedoma; **59**/goodluz; **92**/zstock; **97**/© Peter J. Wilson; **105**/yampi; **107**/warmer; **183**/LingHK; **185**/ArliftAtoz2205; **187**/Aleksandar Todorovic; **187**/Alexander Chaikin; **187**/Andrey Bayda; **187**/Anki Hoglund; **187**/chuyu; **187**/Francesco R. Iacomino; **187**/haveseen; **187**/Jane Rix; **187**/JeniFoto; **187**/Kenneth Dedeu; **187**/leoks; **187**/Mohamed Shareef; **187**/Nickolay Vinokurov; **187**/stocker1970; **187**/Vladimir Melnik; **195**/alexnika; **223**/Radiokafka; **20** (top right)/Dmitriy Kuzmichev; **23** (top right)/nostal6ie; **31** (right)/Keith Wheatley; **41** (top)/Nikola Bilic; **42** (top)/Gimas; **78** (bottom)/Jiratsung; **78** (top left)/Nils Versemann; **78** (top right)/Martin Valigursky; **129** (bottom)/Ryszard Stelmachowicz; **130** (bottom left)/Jill Battaglia; **144** (top left)/joyfull; **155** (bottom left)/T photography; **155** (top right)/Anton_Ivanov; **160** (a)/qingqing; **160** (b)/Ivonne Wierink; **160** (d)/AJP; **160** (c)/NigelSpiers; **166** (top)/John Carnemolla; **169** (bottom left)/edella; **173** (top)/yampi; **181** (bottom)/sigurcamp; **182** (bottom)/Dan Breckwoldt; **182** (top)/Production Perig.com; **199** (top left)/Julia Kuznetsova; **199** (top right)/Jacob Lund; **224** (Alex)/Blend Images; **224** (John)/Sveta Yaroshuk; **224** (Raul)/studioflara.com; **226** (right)/Martin Allinger; **232** (a)/Nils Versemann; **250** (bottom)/pbk-pg; **252** (bottom)/monticello; **256** (bottom)/GagliardiImages; • Spatial Vision: **5**, **32**, **52**, **70**, **86**, **87**, **96**, **114**, **119**, **158**, **167**, **168**, **172**, **180**, **197**, **209**, **237**, **240**, **12** (bottom), **21** (bottom), **38** (bottom), **149** (bottom), **165** (top), **256** (top); **239**/Natural Earth Data; **91**/Base Data © Copyright Commonwealth of Australia Geoscience Australia 2006; **39**, **4** (top)/Geophysical Fluid Dynamics Laboratory, National Oceanic and Atmospheric Administration; **165** (bottom), **201** (top)/Geoscience Australia; **4** (middle right)/NASA Earth Observatory; **62**, **206**, **102** (top)/Natural Earth Data; **4** (bottom left)/USAID, FEWS NET 2011; **31** (left)/Geoscience Australia; **18**/Department of Environment and Water Resources • State Library of NSW: **179**/Mitchell Library, State Library of New South Wales, M3 804eca/1788/1 • State Library of Victoria: **188**, **200**/Pictures Collection; **159** (top left)/John T Collins, Walkerville Lime Kilns, Pictures Collection • Stephen Locke: **110**/Stephen Locke • University of the South Pacific: **247** (c)/Photo as taken by L.Limalevu PACE-SD • World Health Organization: **56**/Safely managed drinking water -thematic report on drinking water 2017. Geneva, Switzerland: World Health Organization; 2017. Licence: CC BY-NC-SA 3.0 IGO • World Resources Institute: **55** (top)/World Resources Institute • World Water Footprint: **66**/WaterFootprint.org and WWF

Text

• ARCADIS: **234**/© Arcadis Sustainable Cities Index 2018 • Australian Bureau of Statistics: **191** • Economist Intelligence Unit: **228**/Data reused by permission of The Economist Intelligence Unit. • John Wiley & Sons Australia, Ltd: **133**; • National Oceanic & Atmospheric Administration: **145** (bottom); **140** (top)/Data from NOAA, drawn by Spatial Vision

Every effort has been made to trace the ownership of copyright material. Information that will enable the publisher to rectify any error or omission in subsequent reprints will be welcome. In such cases, please contact the Permissions Section of John Wiley & Sons Australia, Ltd.

1 Geographical skills and concepts

1.1 Overview

1.1.1 What is Geography?

The world around us is made up of a large range of interesting places, people, cultures and environments. Geography is a way of exploring, analysing and understanding this world of ours: especially its people and places. Studying Geography at school allows you to build up your knowledge and understanding of our planet, at different scales: the local area, our nation, our region and our world. In essence, geographers investigate the characteristics of places and the relationships between people and places.

 **Resources**

✓ **eWorkbook** Customisable worksheets for this topic

LEARNING SEQUENCE

To access a pre-test and starter questions and receive immediate, **corrective feedback** and **sample responses** to every question select your learnON format at www.jacplus.com.au.

1.2 The world of Geography

1.2.1 Geography is ... about our wonderful world

Have you ever visited a place other than the one you live in? If so, you probably would have noticed that some of the features and characteristics are similar and some are different. Geographers aim to understand these characteristics as well as the relationship between people and the different environments around us.

As a geographer, you answer questions ranging from the local to the global, in the past, present and future. Along the way you will develop skills and inquiry methods to answer these questions for yourself.

1.2.2 Geography is ... something you do

One of the best parts of studying Geography is the opportunity to visit places outside the classroom. Going on a field trip allows you to collect data and information for yourself and to work collaboratively with other members of your class.

Geographers use what is called an 'inquiry' approach. This means that you will investigate geographical questions by collecting, analysing, and interpreting information and data in order to develop your own understanding and draw your own conclusions. This helps you develop proposals for what should happen and what action should be taken in the future.

Studying Geography develops a wide range of skills that you can apply in your everyday life, in your future life and possibly in your career!

FIGURE 1 Using maps to work out locations and to plot data

FIGURE 2 Conducting a survey in the field

FIGURE 3 Collecting your own data and information

1.2.3 Geography is ... a way of thinking

Geography is a way of thinking and a way of looking at the world. One of the key tools geographers use is a map. If you look really carefully at them, maps (such as the ones on this page and the next) contain a lot of information. As a student you will often use a variety of different types of maps produced by someone else (e.g. from this textbook, atlases and online). However, as a geographer you will produce your own maps and spatial information, by hand or digitally. Using and interpreting maps are important skills you will develop. It is also important to identify major patterns and trends in maps in order to unlock information they contain.

As a geographer you will use a set of geographical concepts to not only help you think geographically but also to investigate and understand the world. These concepts are *space*, *place*, *interconnection*, *change*, *environment*, *sustainability* and *scale*.

As a geographer you should also ask yourself: 'What can I do and contribute as an informed and responsible citizen to make this world a better place?'

FIGURE 4 Maps: a key tool for the geographer

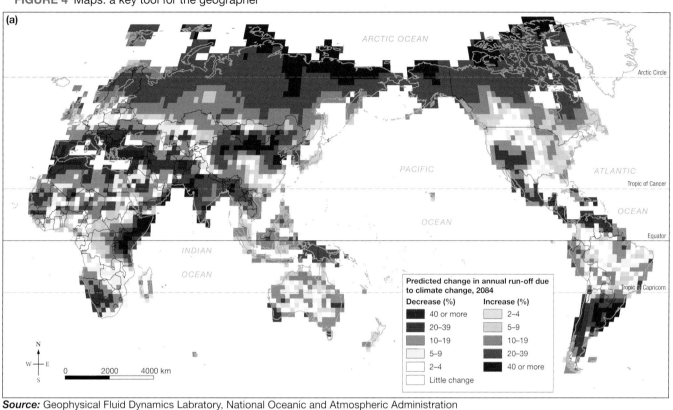

(a)

Predicted change in annual run-off due to climate change, 2084

Decrease (%)	Increase (%)
40 or more	2–4
20–39	5–9
10–19	10–19
5–9	20–39
2–4	40 or more
Little change	

Source: Geophysical Fluid Dynamics Labratory, National Oceanic and Atmospheric Administration

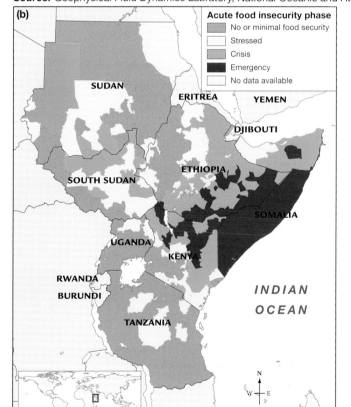

(b)

Acute food insecurity phase
- No or minimal food security
- Stressed
- Crisis
- Emergency
- No data available

Source: USAID, FEWS NET 2011

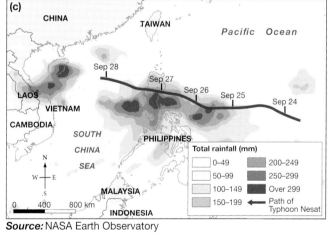

(c)

Total rainfall (mm)

0–49	200–249
50–99	250–299
100–149	Over 299
150–199	← Path of Typhoon Nesat

Source: NASA Earth Observatory

(d)

FIGURE 5 Topographic maps are very useful for geographers as they provide a large amount of detail about places and environments.

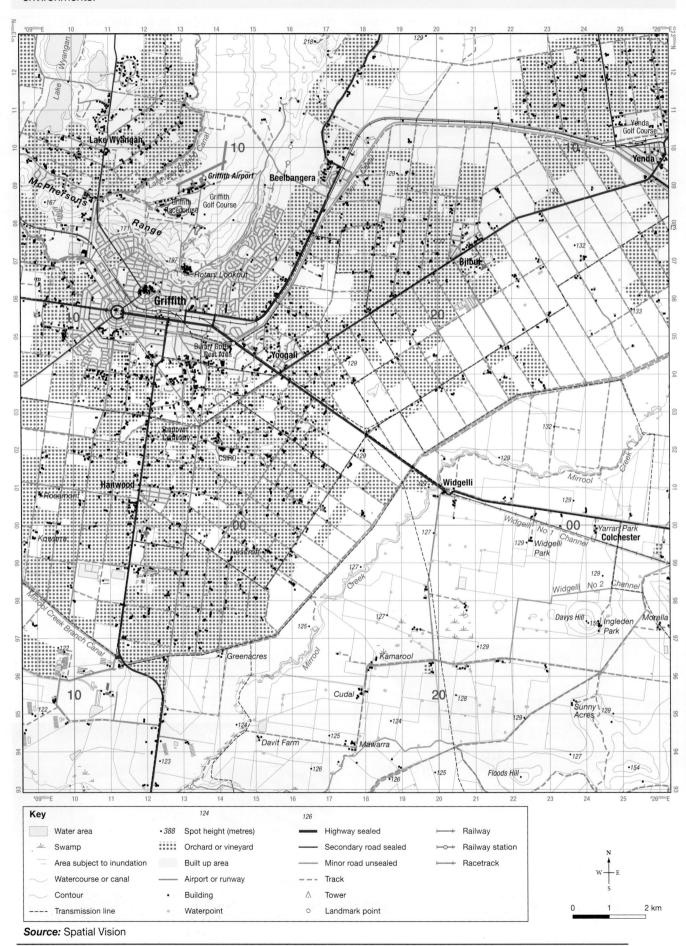

Key

Water area	•388 Spot height (metres)	Highway sealed	Railway
Swamp	Orchard or vineyard	Secondary road sealed	Railway station
Area subject to inundation	Built up area	Minor road unsealed	Racetrack
Watercourse or canal	Airport or runway	Track	
Contour	• Building	∧ Tower	
Transmission line	Waterpoint	○ Landmark point	

0   1   2 km

Source: Spatial Vision

1.3 Concepts and skills used in Geography

1.3.1 Skills used in studying Geography

As you work through each of the topics in this title, you'll complete a range of exercises to check and apply your understanding of concepts covered. In each of these exercises, you'll use a variety of skills, which are identified using the Geographical skills (GS) key provided at the start of each exercise set. These skills are:

- **GS1** Remembering and understanding
- **GS2** Describing and explaining
- **GS3** Comparing and contrasting
- **GS4** Classifying, organising, constructing
- **GS5** Examining, analysing, interpreting
- **GS6** Evaluating, predicting, proposing

In addition to these broad skills, there is a range of essential practical skills that you will learn, practise and master as you study Geography. The SkillBuilder subtopics found throughout this title will tell you about the skill, show you the skill and let you apply the skill to the topics covered.

The SkillBuilders you'll use in Year 7 are listed below.

- Constructing a pie graph
- Annotating a photograph
- How to read a map
- Drawing a line graph
- Reading a weather map
- Interpreting diagrams
- Cardinal points: wind roses

- Creating a simple column or bar graph
- Using topographic maps
- Creating a concept diagram
- Understanding satellite images
- Using alphanumeric grid references
- Drawing a climate graph
- Creating and analysing overlay maps

1.3.2 SPICESS

Geographical concepts help you to make sense of your world. By using these concepts you can both investigate and understand the world you live in, and you can use them to try to imagine a different world. The concepts help you to think geographically. There are seven major concepts: ***space***, ***place***, ***interconnection***, ***change***, ***environment***, ***sustainability*** and ***scale***. You will use the seven concepts to investigate two units: *Water in the world* and *Place and liveability*.

FIGURE 1 A way to remember these seven concepts is to think of the term SPICESS.

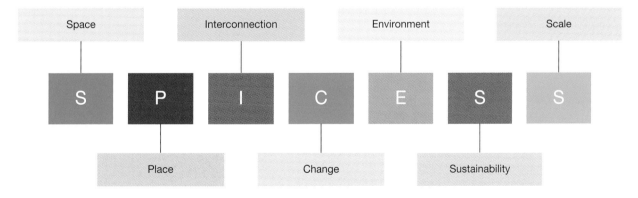

1.3.3 What is space?

Everything has a location on the space that is the surface of the Earth, and studying the effects of location, the distribution of things across this space, and how the space is organised and managed by people, helps us to understand why the world is like it is.

A place can be described by its absolute location (latitude and longitude) or its relative location (in what direction and how far it is from another place).

FIGURE 2 Australian annual rainfall variability

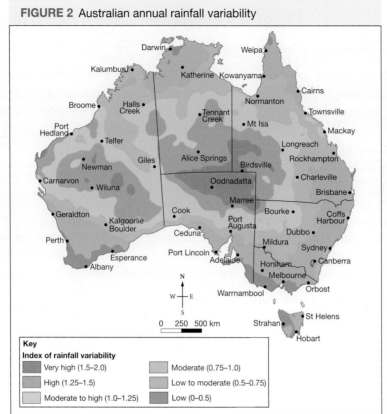

Source: MAPgraphics Pty Ltd, Brisbane

FIGURE 3 The amount of rain that falls in Australia varies from place to place, as this rainfall map shows.

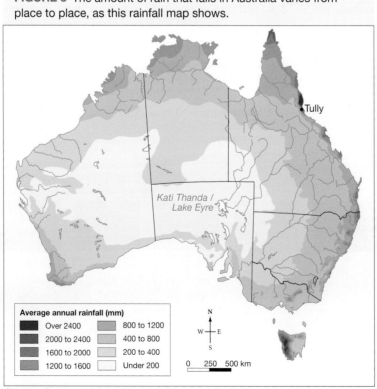

Source: Bureau of Meteorology, 2003, on the Australian Water Map, Earth Systems Pty Ltd

1.3.4 What is place?

The world is made up of places, so to understand our world we need to understand its places by studying their variety, how they influence our lives and how we create and change them.

You often have mental images and perceptions of places — your city, suburb, town or neighbourhood — and these may be very different from someone else's perceptions of the same places.

FIGURE 4 Mount Tom Price township and mine in Western Australia, with fly in, fly out (FIFO) workers huts in the left foreground

1.3.5 What is interconnection?

People and things are connected to other people and things in their own and other places, and understanding these connections helps us to understand how and why places are changing.

An event in one location can lead to change in a place some distance away.

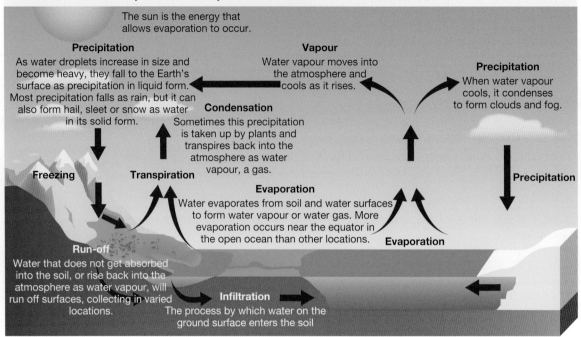

FIGURE 5 The water cycle shows many interconnections.

The sun is the energy that allows evaporation to occur.

Precipitation
As water droplets increase in size and become heavy, they fall to the Earth's surface as precipitation in liquid form. Most precipitation falls as rain, but it can also form hail, sleet or snow as water in its solid form.

Vapour
Water vapour moves into the atmosphere and cools as it rises.

Precipitation
When water vapour cools, it condenses to form clouds and fog.

Condensation
Sometimes this precipitation is taken up by plants and transpires back into the atmosphere as water vapour, a gas.

Freezing

Transpiration

Precipitation

Evaporation
Water evaporates from soil and water surfaces to form water vapour or water gas. More evaporation occurs near the equator in the open ocean than other locations.

Evaporation

Run-off
Water that does not get absorbed into the soil, or rise back into the atmosphere as water vapour, will run off surfaces, collecting in varied locations.

Infiltration
The process by which water on the ground surface enters the soil

---

Explore more with my**World**Atlas

Deepen your understanding of this topic with related case studies and questions.
• Developing Australian Curriculum concepts > **Interconnection**

1.3.6 What is change?

The concept of change is about using time to better understand a place, an environment, a spatial pattern or a geographical problem.

The concept of change involves both time and space — change can take place over a period of time, or over an area. The time period for change can be very short (for example, the impact of a flash flood) or over thousands or millions of years (for example, the development of fossil fuel resources).

FIGURE 6 Port Douglas, 60 km north of Cairns, was a busy port in the 1870s, with a population over 10 000. The mining that had attracted people to this hot, wet area did not last. By the 1960s, the population was only 100. In the 1980s, road and air access to the town improved and tourist numbers to the area grew. The permanent population is now about 3500. During the peak holiday season (May to November), this number increases by four times.

(a) 1971

(b) 2019

Environmental change can occur over short or long periods of time. The use of technology can result in rapid change — think of the explosions at a mining site that reveal mineral seams.

The degree of change occurring can be used to predict, or plan for, actual or preferred futures.

Explore more with myWorldAtlas

Deepen your understanding of this topic with related case studies and questions.
• Developing Australian Curriculum concepts > **Change**

1.3.7 What is environment?

People live in and depend on the environment, so it has an important influence on our lives.

The environment, defined as the physical and biological world around us, supports and enriches human and other life by providing raw materials and food, absorbing and recycling wastes, and being a source of enjoyment and inspiration to people.

FIGURE 7 Pacific Islanders use traditional methods to fish sustainably.

Explore more with myWorldAtlas

Deepen your understanding of this topic with related case studies and questions.
• Developing Australian Curriculum concepts > **Environment**

1.3.8 What is sustainability?

Sustainability is about maintaining the capacity of the environment to support our lives and those of other living creatures.

Sustainability is about the interconnection between the human and natural world and who gets which resources and where, in relation to conservation of these resources and prevention of environmental damage.

FIGURE 8 The process of shifting cultivation means that farmers move on when an area becomes unproductive, allowing the land to recover.

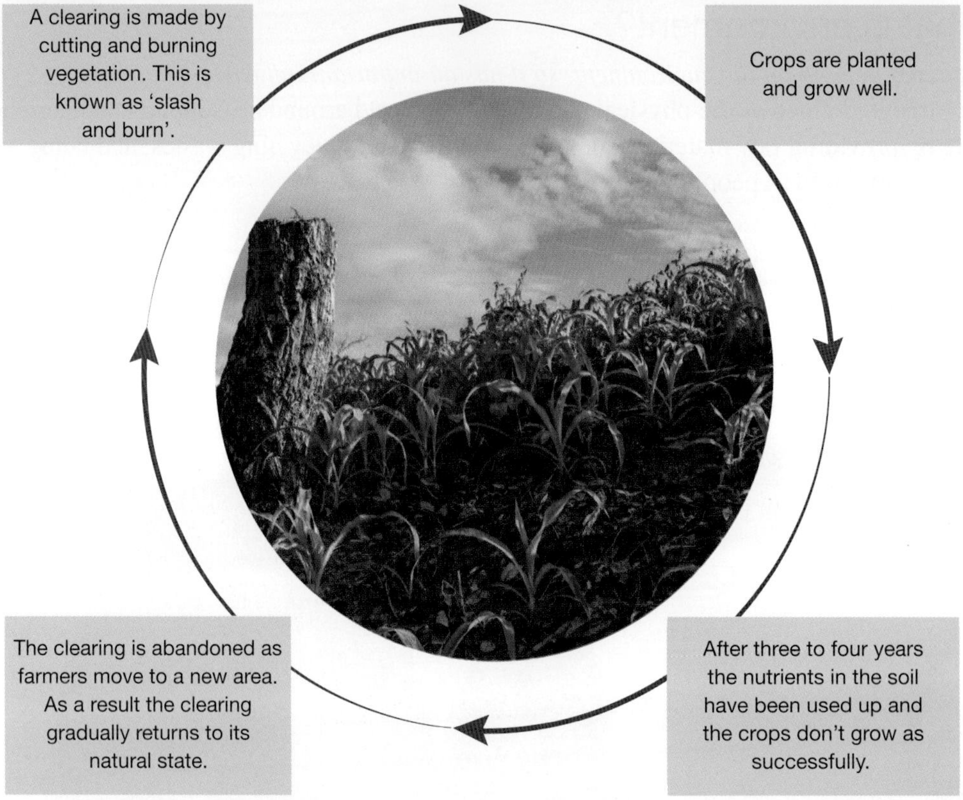

A clearing is made by cutting and burning vegetation. This is known as 'slash and burn'.

Crops are planted and grow well.

The clearing is abandoned as farmers move to a new area. As a result the clearing gradually returns to its natural state.

After three to four years the nutrients in the soil have been used up and the crops don't grow as successfully.

Explore more with my**World**Atlas

Deepen your understanding of this topic with related case studies and questions.
• Developing Australian Curriculum concepts > **Sustainability**

1.3.9 What is scale?

When we examine geographical questions at different spatial levels we are using the concept of scale to find more complete answers.

Scale can be applied at personal and local levels to regional, national or global levels. Looking at things at a range of scales allows a deeper understanding of geographical issues.

Different factors can be involved in explaining phenomena at different scales. Local events can have global outcomes; for example, removing areas of forest at a local scale can have an impact on climate at a global scale. A policy at a national scale, such as forest protection, can have an impact at a local scale, such as the protection of an endangered species.

FIGURE 9 Mental map of Jayden's local place (a) by Jayden and (b) by Annette, Jayden's mother

(a)

(b)

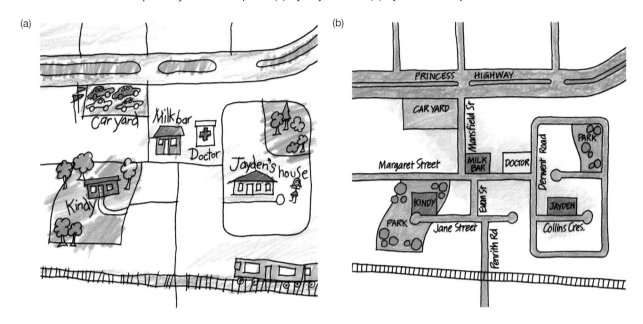

FIGURE 10 Railway route and main settlements between Sydney and Perth

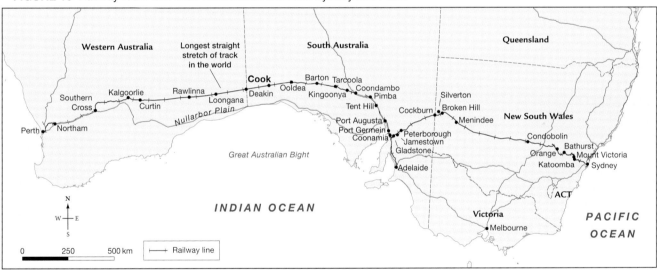

Source: Spatial Vision

Explore more with my**World**Atlas

Deepen your understanding of this topic with related case studies and questions.
• Developing Australian Curriculum concepts > **Scale**

1.4 Review

1.4.1 Key knowledge summary

Use this dot point summary to review the content covered in this topic.

1.4 Exercise 1: Review

Select your learnON format to complete review questions for this topic.

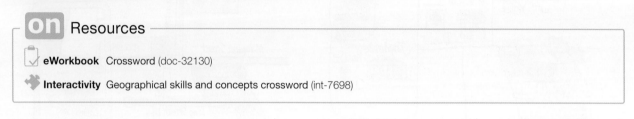

UNIT 1
WATER IN THE WORLD

Water is all around us — in the atmosphere as clouds; as snow on mountain tops; in rivers, lakes and dams on the Earth's surface; in our swimming pools, on our farms and in our taps. It is one of our most valuable resources, because humans cannot survive without water. Therefore, we need to learn to use and look after this resource very carefully.

FIELDWORK INQUIRY: WHAT IS THE WATER QUALITY OF MY LOCAL CATCHMENT?

Task

Your team has been selected to research the water quality of a local catchment or waterway and produce a report and presentation on your findings. Be sure to measure water quality at different locations along the river, creek or stream, and try to determine the causes of different water quality.

Select your learnON format to access:
- an overview of the project task
- details of the inquiry process
- resources to guide your inquiry
- an assessment rubric.

 Resources

ProjectsPLUS Fieldwork inquiry: What is the water quality of my local catchment? (pro-0143)

2 Our precious environmental resources

2.1 Overview

Earth provides us with more than a place to live. What resources do we take from it to survive?

2.1.1 Introduction

Have you ever stopped to think about the resources you need to survive every day? Fortunately, the Earth supplies us with the natural resources we need for our food, shelter, clothing, and energy for our homes and factories. These resources include water, fossil fuels and mineral deposits. However, access to these supplies is not distributed equally around the planet, and attitudes towards them may differ or change over time. As the global population increases, great damage is being done to the environment as a result of using these resources. Moreover, we need to carefully manage our use of them to ensure that these resources are available for use in the future.

---

on Resources

☑ **eWorkbook**   Customisable worksheets for this topic

🎞 **Video eLesson** Water: A vital resource (eles-1615)

LEARNING SEQUENCE

2.1 Overview
2.2 Environmental resources
2.3 Non-renewable energy in Australia and the world
2.4 **SkillBuilder:** Constructing a pie graph   `online only`
2.5 Water as a resource
2.6 **SkillBuilder:** Annotating a photograph   `online only`
2.7 Groundwater as a resource
2.8 **Thinking Big research project:** The Great Artesian Basin   `online only`
2.9 **Review**   `online only`

To access a pre-test and starter questions and receive immediate, **corrective feedback** and **sample responses** to every question, select your learnON format at www.jacplus.com.au.

2.2 Environmental resources

2.2.1 Why do we need resources?

We depend on environmental resources to survive. We need water to drink, soil to produce our food, and forests and mines to supply other materials. Environmental resources (also called natural resources) are raw materials that occur in the environment and which are necessary or useful to people. They include soil, water, mineral deposits, **fossil fuels**, plants and animals.

Think about all the resources you have used today from the time you woke up until the time you reached the school gate. Perhaps you used water to shower, brush your teeth, wash the dishes or as a refreshing drink? Consider all the different foods that had to be farmed to provide the ingredients for your breakfast. Finally, how did you get to school? If you used a form of transport, there is a good chance a resource powered it!

There are two types of environmental resources: renewable and non-renewable. Renewable resources are those that can be replaced in a short time. For example, solar energy is a renewable resource that can be used for heating water or generating electricity. It is never used up and is constantly being replaced by the sun. Non-renewable resources are those that cannot be replaced in a short time. For example, fossil fuels such as oil, coal and natural gas are non-renewable because they take thousands of years to be replaced.

We cannot make more non-renewable resources; they are limited and will eventually run out. However, renewable environmental resources are things that can grow and be replaced over time if they are carefully managed. Forests, soils and fresh water are renewable.

FIGURE 1 Many resources are required to provide a family with breakfast.

FIGURE 2 Environmental resources — renewable and non-renewable

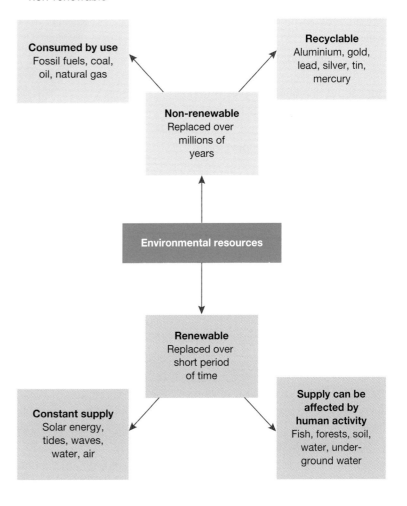

Consumed by use
Fossil fuels, coal, oil, natural gas

Recyclable
Aluminium, gold, lead, silver, tin, mercury

Non-renewable
Replaced over millions of years

Environmental resources

Renewable
Replaced over short period of time

Constant supply
Solar energy, tides, waves, water, air

Supply can be affected by human activity
Fish, forests, soil, water, underground water

2.2.2 Global supply

The global distribution of environmental resources depends on geology (the materials and rocks that make up the Earth) and climate. Some minerals are rare and are found in only a few locations. For example, **uranium** is found mainly in Australia. Several countries in the Middle East, such as Saudi Arabia and Iran, have rich oil resources but are short of water. Many countries in Africa, such as Botswana, have mineral resources but lack the money to mine and process them.

The human activities of agriculture, fishing, logging and mining all depend directly on natural resources. In developing countries, traditional forms of agriculture such as **subsistence farming** and nomadic herding are still common. These activities are sustainable if farmers move on when an area becomes unproductive, allowing the land to recover. However, poverty and population growth mean that many people now clear forests for farms and overgraze or over crop small plots of land, resulting in deforestation and land degradation.

Farms in developed countries are usually much larger. For example, the Anna Creek cattle station in South Australia is 24 000 square kilometres, the size of Belgium. In contrast, an average intensive rice farm in Bali is only about one hectare. This is about four times the size of an Australian quarter-acre block of land. Unsustainable agricultural practices in developed countries include the overuse of water, fertilisers and pesticides. For example, fertilisers help crops to grow but, when they end up in rivers and oceans as run-off, they cause algal blooms and damage coral reefs.

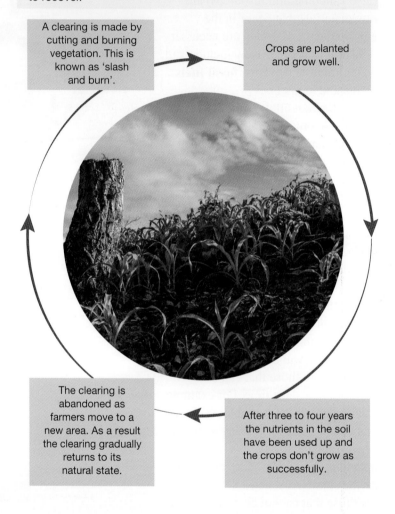

FIGURE 3 The process of shifting cultivation means that farmers move on when an area becomes unproductive, allowing the land to recover.

A clearing is made by cutting and burning vegetation. This is known as 'slash and burn'.

Crops are planted and grow well.

After three to four years the nutrients in the soil have been used up and the crops don't grow as successfully.

The clearing is abandoned as farmers move to a new area. As a result the clearing gradually returns to its natural state.

2.2.3 Australia's environmental resources

Australia has more environmental resources per head of population than any other country in the world. The main reason for this is that we have a small population in a very large country.

2.2.4 Minerals

Australia is rich in mineral resources. The Pilbara region in Western Australia, for example, has some of the largest reserves of iron ore in the world. Australia also produces many other minerals, including silver, copper, nickel and tin.

Australia is the world's:
- largest exporter of iron ore, bauxite, lead, diamonds, zinc ores and mineral sands
- second-largest exporter of alumina (processed from bauxite and then turned into aluminium)
- third-largest exporter of gold.

2.2.5 Soils

Australia has generally poor soils, especially when compared with those found in other continents such as North America and Europe. Most Australian soils are low in nutrients, and in some parts of the continent, particularly the more arid areas, high salt content is also a problem. Most parts of Australia are suitable only for sheep and cattle grazing, rather than **intensive agriculture**, owing to low rainfall and poor soils.

There are regions of good soil scattered throughout Australia. These include soils formed from volcanic rock, such as those on the Darling Downs in Queensland and around Orange in New South Wales, and **alluvial soils** (found in river valleys).

2.2.6 CASE STUDY: The Pilbara

Most of Australia's iron ore reserves are found in the Pilbara region in north-west Western Australia. The Pilbara accounts for 98 per cent of the country's iron ore production and 96 per cent of its exports. Iron ore is the raw material from which iron is made. Although iron in its cast form has many uses, its main use is in steelmaking. Steel is the main structural metal in engineering, building, ship building, cars and machinery.

Two of the largest companies operating in the Pilbara are BHP Billiton and Rio Tinto. There are many mines in the Pilbara, including those at Mount Tom Price, Marandoo, Channar, Newman and Robe River.

The iron ore in the Pilbara is relatively easy to mine. It is also high quality, so there is strong demand for it from many countries, including Japan, China and South Korea.

FIGURE 4 Mineral deposits in the Pilbara region

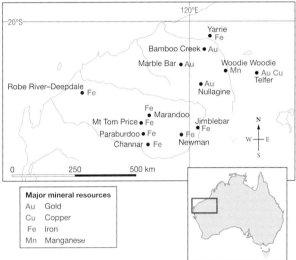

Source: MAPgraphics Pty Ltd, Brisbane

FIGURE 5 The Mount Tom Price iron ore mine in the Pilbara

2.2.7 Natural scenery

Australia's spectacular scenery attracts tourists from all over the world, particularly to those sites that are on the World Heritage List. This means that they are recognised as being of global importance due to their great natural or cultural significance.

2.2.8 Forests

Apart from Antarctica, which has no trees, Australia is the world's least forested continent. The most common vegetation in Australia is woodland and shrubland. Before European occupation, about 9 per cent of Australia was forested. Today, about 5 per cent of the country is forested. Even though Australia exports timber products, it also imports a lot of timber, particularly softwoods such as pine.

FIGURE 6 Australia's World Heritage sites

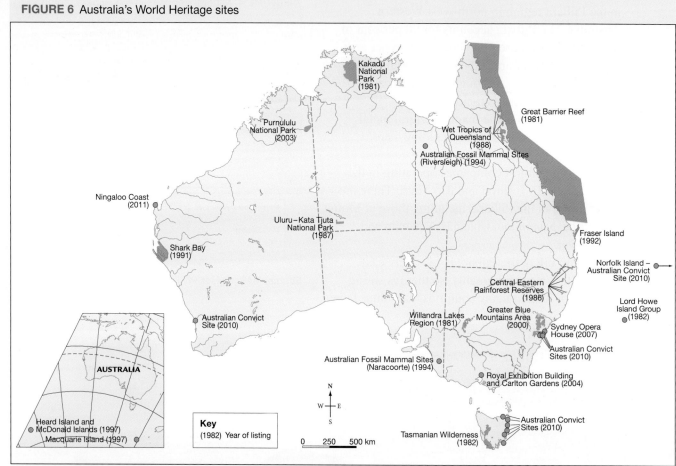

Source: Department of Environment and Water Resources

On Resources

🔗 **Weblink** UNESCO criteria

🛰 **Google Earth** Mt Tom Price Mine

Explore more with myWorldAtlas

Deepen your understanding of this topic with related case studies and questions.
* Exploring places > The world > **World Heritage sites**

2.2 INQUIRY ACTIVITIES

1. Refer to **FIGURE 3**. Use the internet to research who uses shifting cultivation around the world. Choose one case study and report back to the class about their way of life. Examples of these may include tribes from *places* such as the Amazon, Congo Basin or Papua New Guinea. Compare your chosen tribe's way of life with your way of life, and explain how it differs when it comes to using resources and accessing food. Upon completion of the presentations, discuss as a class why you think Australian farmers do not use shifting cultivation as their method of agricultural production.   **Comparing and contrasting**

2. Create a table that lists ten renewable and ten non-renewable resources used by your family. Be specific; for example, list timber used in your furniture. From your list, note some of the waste and pollution that may be created in the use or creation of these resources. How could this be reduced to improve environmental *sustainability*? As a class develop a five-point plan of how you could all be more proactive in being more *environmentally sustainable* every day.   **Evaluating, predicting, proposing**

3. Using the internet, research the criteria required for an area to become classified as a World Heritage area.
                    Examining, analysing, interpreting

4. In groups of two or three, research one of the World Heritage sites found in Australia. Each group should investigate a different area. Present your group's research to the class and accompany your presentation with a PowerPoint or other graphic presentation so your peers can see the features that make the *place* so special.   **Examining, analysing, interpreting**

2.2 EXERCISES

Geographical skills key: GS1 Remembering and understanding **GS2** Describing and explaining **GS3** Comparing and contrasting **GS4** Classifying, organising, constructing **GS5** Examining, analysing, interpreting **GS6** Evaluating, predicting, proposing

2.2 Exercise 1: Check your understanding

1. **GS1** What is an *environmental* resource?
2. **GS2** Outline the difference between renewable and non-renewable resources.
3. **GS1** List three examples of non-renewable resources that can be recycled.
4. **GS2** Which renewable resources are most affected by human activity? Why?
5. **GS1** List the main mineral resources produced in Australia.
6. **GS5** Refer to **FIGURE 6**. Which state of Australia has the most World Heritage sites?
7. **GS2** Why are many Australian soils suitable only for grazing?
8. **GS2** Why does Australia's scenery attract many overseas visitors?

2.2 Exercise 2: Apply your understanding

1. **GS6** When it comes to using *environmental* resources, there are two main problems people face. What are they and why are they important?
2. **GS2** What does the *sustainable* use of *environmental* resources mean?
3. **GS2** Why is shifting cultivation a *sustainable* form of agriculture?
4. **GS5** Refer to the case study on the Pilbara.
   (a) Where does Australia's main iron ore production take place?
   (b) What is the main use of iron ore?
   (c) Why is there strong demand overseas for Australian iron ore?
   (d) List three *places* in the Pilbara where there are deposits of iron ore.
5. **GS1** What are examples of fossil fuels that you use in order to maintain your lifestyle?
6. **GS5** Refer to **FIGURE 4**. If you are at the Mount Tom Price mine, in which direction and how far are the following *places*?
   (a) Yarrie
   (b) Robe River–Deepdale
   (c) Telfer
   (d) Channar
   What is mined at each location?

Try these questions in learnON for instant, corrective feedback. Go to www.jacplus.com.au.

2.3 Non-renewable energy in Australia and the world

2.3.1 Australia's energy use

Australia has large reserves of non-renewable energy, such as coal, natural gas, oil and uranium. Over the past 30 years, Australia's energy consumption has increased by over 200 per cent. Most of the energy we use comes from non-renewable sources, particularly coal, which is used for steel manufacturing and 73 per cent of electricity generation.

Each year, over half of Australia's energy products are exported. Australia is the world's fourth largest producer of coal. In the past five years, coal has made up approximately 15 per cent of Australia's exports. Demand for coal from Asia has increased over the past ten years. Japan is our largest market, claiming 45 per cent of Australian coal. The next biggest importer of Australian coal is China, which takes 23 per cent of the market. There is the possibility that demand for Australian coal from South-East Asia will triple in the next 25 years.

FIGURE 1 The biggest buyers in the Australian coal market, 2018

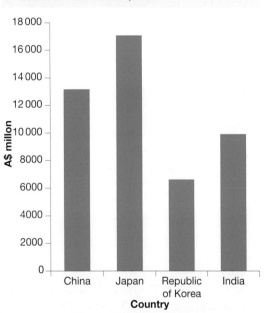

Source: Department of Foreign Affairs and Trade

FIGURE 2 Extraction of coal carries a high environmental cost.

FIGURE 3 Millions of tonnes of coal are moved around Australia annually.

FIGURE 4 Australian non-renewable energy consumption by fuel

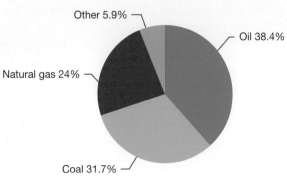

2.3.2 Global use

The world relies heavily on energy for transport, heating and manufacturing. The amount of energy used varies widely around the world. The world's most commonly used energy sources are oil, coal and gas (all fossil fuels), hydro-electricity and nuclear power.

Internationally, there is more trade in oil than any other product. Oil is unevenly distributed, with most reserves located in the Middle East. As a result, other nations need to buy their oil from this region.

Oil is a fossil fuel and a non-renewable resource. It is believed that oil reserves will eventually run out, probably within 40 to 80 years. Between half and two-thirds of oil production is used in transport. It is also used to produce energy and to **manufacture** products such as plastic, nail polish, lipstick, synthetic textiles and whitegoods.

Every day, 92 million barrels of oil are used around the world; the United States uses 19 million barrels per day while the second highest consumer of oil, China, uses 11 million barrels per day. Australia consumes almost 1 million (or 998 thousand) barrels per day.

Oil is exported to countries that can afford to pay for it. For example, even though the United States has only 5 per cent of the world's population, it consumes around 20 per cent of global supplies. Some countries import oil because they use more than they produce.

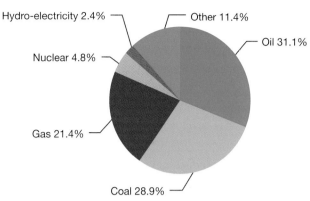

FIGURE 5 World energy production

Hydro-electricity 2.4%
Other 11.4%
Oil 31.1%
Nuclear 4.8%
Gas 21.4%
Coal 28.9%

FIGURE 6 Global oil consumption

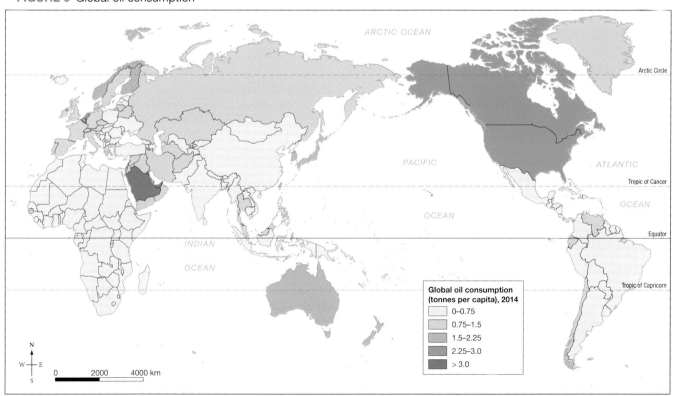

Global oil consumption
(tonnes per capita), 2014
- 0–0.75
- 0.75–1.5
- 1.5–2.25
- 2.25–3.0
- > 3.0

Source: Spatial Vision

2.3.3 Fossil fuel winners and losers

Governments, oil companies and individuals make billions of dollars from oil. For example, Saudi Arabia's Crown Prince and the Sultan of Brunei are oil billionaires, and large oil companies are some of the most profitable companies in the world.

Oil generates economic growth and can improve people's living standards in producer countries. However, while oil may bring wealth to governments and corporations, wealth does not always trickle down to local populations. Venezuela is one of the world's top ten oil producers, yet 32 per cent of its population lives below the poverty line.

2.3.4 Australia's renewable energy

Australia has many potential renewable energy sources: hydro-electric, wind, solar, tidal, geothermal and **biomass** energy. Although the use of these is increasing, renewables provide us with only about 6 per cent of our energy needs. In 2015, the Australian government reviewed its Renewable Energy Target and set the goal that by 2020 at least 33 000 gigawatt-hours, or 23.5 per cent of Australia's electricity, would come from renewable sources. This would be enough electricity to power approximately five million homes for a year. Other benefits may include billions of dollars of investment, the creation over 15 000 jobs and, in the long term, it could even save every Australian household $140 off their electricity bill annually.

Australia is a large country with a number of different environments that are suited to supplying renewable energy. There has been an increase in public and government concern over global warming and climate change, so more research into and development of renewable energy resources may help to minimise these concerns.

FIGURE 7 Australia's renewable energy consumption, by fuel type

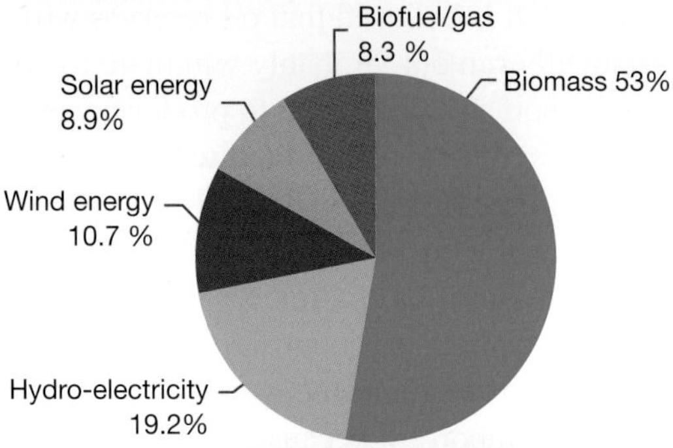

- Biofuel/gas 8.3 %
- Solar energy 8.9%
- Biomass 53%
- Wind energy 10.7 %
- Hydro-electricity 19.2%

Note: Totals may not add due to rounding.

DISCUSS

Hydro-electricity is a form of renewable energy. Should Australian states and territories be encouraged to build more dams on major rivers to develop renewable hydro-electric power generation? Conduct a class discussion from various viewpoints including:
- Indigenous Australian groups who own land being flooded
- the government
- hydro-electrical engineers
- environmentalists.

[Personal and Social Capability]

2.3.5 Different types of energy

Some types of renewable energy that could be sustainable in Australia include the following.
- *Geothermal energy.* Australia has huge underground energy resources known as 'hot rocks'. Water can be heated by pumping it underground through these hot rocks. The resulting steam then drives a **turbine** to generate electricity.
- *Wind power.* A typical wind turbine can meet the energy needs of up to 1000 homes.
- *Tidal power.* Waves can drive turbines to produce electricity. Tidal power is especially suitable for powering **desalination** plants, though Australia does not use this system.
- *Biomass energy.* This is produced from the combustion of organic matter, such as sugar cane and corn crops. Biomass can be used to produce electricity as well as liquid fuels like ethanol and biodiesel.

- *Solar power*. Solar energy technologies harness the sun's heat and light to provide heating, lighting and electricity. Two types of solar technologies are currently under development in Australia.
  - Photovoltaic cells convert solar energy directly into electricity. These can be placed on roofs in order to collect direct sunlight.
  - Solar thermal systems use the sun's heat to generate electricity by first heating a fluid such as water to create steam, which drives a turbine to generate electricity. Australia has abundant solar radiation, and therefore great potential for the development of solar energy. Germany is currently the highest consumer of solar electricity in the world.
- *Hydro-electric power*. Most of Australia's hydro-electric power is generated in Tasmania and by the Snowy Mountains Hydro-Electric Scheme in New South Wales. Around 6 per cent of Australia's electricity comes from hydro-electric power, but there are limited opportunities for increasing this because of the lack of water resources, and because building dams can be controversial.

The use of sustainable energy is still in its early stages but is growing rapidly in China, the United States and Europe. Fossil fuels are currently cheaper and more convenient to produce than renewable energy sources. The reason for this is that we do not pay the real cost of their use — we do not pay for the huge cost of releasing waste products into the **atmosphere**. In future, a carbon tax or restrictions on the use of fossil fuels would increase their cost, perhaps making renewable energy a more attractive option for consumers.

FIGURE 8 Some sources of renewable energy: (a) solar, (b) biomass, (c) wind, (d) hydro-electric (e) geothermal, (f) tidal

─ Explore more with my**World**Atlas ──────────────────────────────

Deepen your understanding of this topic with related case studies and questions.
- Investigate additional topics > Energy > **World energy**
- Investigate additional topics > Energy > **Energy in Australia**

on Resources

🔗 **Weblinks**   Gulf of Mexico

NOAA

Treehugger

2.3 INQUIRY ACTIVITY

The extraction of fossil fuels can have negative impacts on the *environment*. Using the internet, research the 2010 BP oil disaster in the Gulf of Mexico. Create a poster that illustrates the causes of the event and its effects, and explains the clean-up methods used to try to contain the damage. Ensure your presentation is visually appealing by using a variety of maps, photos and graphs.   **Examining, analysing, interpreting**

2.3 EXERCISES

Geographical skills key: GS1 Remembering and understanding **GS2** Describing and explaining **GS3** Comparing and contrasting **GS4** Classifying, organising, constructing **GS5** Examining, analysing, interpreting **GS6** Evaluating, predicting, proposing

2.3 Exercise 1: Check your understanding

1. **GS1** Refer to **FIGURE 1**.
   (a) In which geographic region are these nations located?
   (b) How many more per cent of coal does China import than Taiwan?
2. **GS1** Other than energy, what other products are manufactured from oil?
3. **GS1** How many more barrels of oil does the United States use than Australia per day?
4. **GS2** Why doesn't the money made from the sale of fossil fuels, such as oil, always benefit the local people in the country that sells the product?
5. **GS1** In which renewable energy resources is Australia (a) rich and (b) poor?
6. **GS1** Define the term 'renewable energy'. List the main forms of renewable energy.
7. **GS1** What are the two types of solar technology?

2.3 Exercise 2: Apply your understanding

1. **GS2** Why is renewable energy not widely used in many countries?
2. **GS6** What will happen to Australian coal reserves if demand from other countries continues to grow? What could be some of the impacts of this? Does this raise any ethical dilemmas?
3. **GS5** Refer to the pie graph in **FIGURE 7**. What type of renewable resource provides the greatest source of energy in Australia?
4. **GS6** Australia is a sunny country, so why does only 8.9 per cent of our country's renewable energy come from the sun?
5. **GS6** List some of the likely positive and negative consequences if Australia stopped using non-renewable energy resources altogether and replaced them with renewable energy sources.

Try these questions in learnON for instant, corrective feedback. Go to www.jacplus.com.au.

2.4 SkillBuilder: Constructing a pie graph

What is a pie graph?

A pie chart, or pie graph, is a graph in which slices or segments represent the size of different parts that make up the whole. The size of the segments is easily seen and can be compared. Pie graphs give us an overall impression of data.

Select your learnON format to access:

- an overview of the skill and its application in Geography (Tell me)
- a video and a step-by-step process to explain the skill (Show me)
- an activity and interactivity for you to practise the skill (Let me do it)
- questions to consolidate your understanding of the skill.

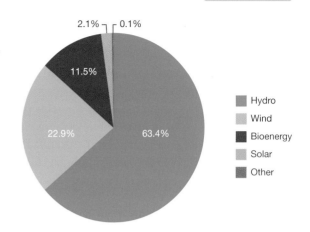

2.1% — 0.1%

11.5%

22.9%   63.4%

- Hydro
- Wind
- Bioenergy
- Solar
- Other

on Resources

Video eLesson   Constructing a pie graph (eles-1632)

Interactivity   Constructing a pie graph (int-3128)

2.5 Water as a resource

2.5.1 Is water a renewable resource?

The amount of water on Earth has not changed since the beginning of time; there is only a finite, or fixed, amount. The water used by ancient and extinct animals and plants millions of years ago is the same water that today falls as rain. The amount of water cannot be increased or decreased. It is cycled and recycled, and constantly changes its state from gas, to liquid, to solid, and back.

Water is a resource. Like any other resource, it has no value in itself, but has great value when a use is found for it. Environments where water is found are also a resource. A river can be the site of a settlement that provides transport as well as food. A **riverine environment** that includes fish, birds, wildlife, wetlands, plants and micro-organisms is also valuable as a living system and can therefore be regarded as a resource.

Water is available in different forms and sources. Green water is described as water from precipitation that is stored in the root zone of the soil and transpired or used by plants. Green water is used in many forms of agricultural, horticulture and forestry that rely on precipitation to grow products.

Blue water is water that has been sourced from surface or groundwater. It is either evaporated, incorporated into a product or taken from one body of water (e.g. a river) and returned to another (e.g. an irrigation channel). Irrigated agriculture, industry and domestic water use blue water.

2.5.2 The water cycle

All the water on Earth moves through a cycle that is powered by the sun. This cycle is called the water cycle, or **hydrologic cycle**. Water is constantly changing its location (through constant movement) and its form (from gas, to liquid, to solid). Evaporation, condensation and freezing of water occur during the cycle.

It is estimated that up to 70 per cent of the Earth's water is locked in ice sheets in the Arctic, Greenland and Antarctica. This water is therefore a potential resource — it exists in a location and in a form in which it cannot be immediately used. Water is also a potential resource when it exists as a gas (water vapour), as salt water or as waste water.

The water cycle shows how water connects places. Water flows through the environment in different forms. Because it is such an essential resource for life, places are closely connected to water and water supply. Many places are settled on rivers and lakes with access to water. Rivers connect places and agriculture is connected to water supply (surface or groundwater).

FIGURE 1 The water cycle

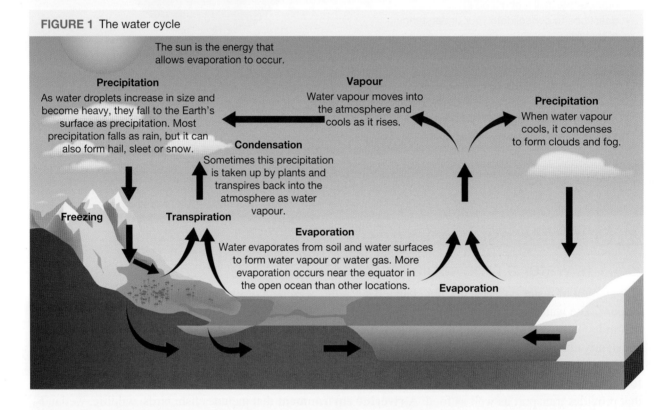

2.5.3 How long does water stay in the one place?

Water can stay in the one place very briefly or it can stay for many thousands of years. It has been calculated that water stays in the atmosphere for an average of nine days before it falls to Earth again as precipitation. Water stays in soil for between one and two months. If you live in an area that has experienced drought or a very long summer without rain, you may have noticed that the soil dries out and forms cracks. Once the seasons change and it begins to rain, the soil absorbs water again and the cracks disappear.

Water spends between two and six months in snow and rivers but a lot longer in large lakes, glaciers, oceans and **groundwater**. The longest time water stays in one place is in the Antarctic ice sheets. Some ice core samples in Antarctica contain water that is 800 000 years old, but the average is about 20 000 years.

The length of time water spends as groundwater can be an average of 10 000 years if it is very deep, but it can stay much longer.

FIGURE 2 A scientist working with ice core samples in Antarctica. Some of the longest records of our climate have come from large ice sheets over three kilometres thick in Greenland and Antarctica. They produce records going back several hundred thousand years.

 Resources

 Interactivity Water works (int-3077)

2.5 INQUIRY ACTIVITY

Using the information about the water cycle, write and present 'The incredible journey of water'. Focus on various **interconnections** between water and the **environment**. Include diagrams and photographs, and present your story electronically, as a poster, a written piece, a drama piece or a song.   **Describing and explaining**

2.5 EXERCISES

Geographical skills key: GS1 Remembering and understanding **GS2** Describing and explaining **GS3** Comparing and contrasting **GS4** Classifying, organising, constructing **GS5** Examining, analysing, interpreting **GS6** Evaluating, predicting, proposing

2.5 Exercise 1: Check your understanding

1. **GS1** Is water a renewable or a non-renewable resource?
2. **GS1** List all the ways that water can be used as a resource.
3. **GS1** Name two **places** where water stays in the same **place** for the longest.
4. **GS2** How does the water cycle prove that we are using the same water that the dinosaurs used — in other words, that it is finite (limited)?
5. **GS2** Explain how the hydrologic cycle moves water across the Earth.

2.5 Exercise 2: Apply your understanding

1. **GS6** With a global population that is increasing by about 75 million people each year, how will it be possible for the finite water on Earth to be shared fairly?
2. **GS2** What are ice core samples and and what sort of information do they provide?   ▶

3. **GS2** How is water vapour related to the process of evaporation?
4. **GS5** How does the process of precipitation interconnect water in the water cycle?
5. **GS2** Describe how a riverine environment interconnects with water.

Try these questions in learnON for instant, corrective feedback. Go to www.jacplus.com.au.

2.6 SkillBuilder: Annotating a photograph

Using annotated photographs in Geography

Photographs are used to show aspects of a place. Annotations are added to photographs to draw the reader's attention to what can be seen and deduced.

Select your learnON format to access:

- an overview of the skill and its application in Geography (Tell me)
- a video and a step-by-step process to explain the skill (Show me)
- an activity and interactivity for you to practise the skill (Let me do it)
- questions to consolidate your understanding of the skill.

Power line
Riparian vegetation
Sandy bank
Snag
Water pump input
River

on Resources

🎞 **Video eLesson**   Annotating a photograph (eles-1633)

🧩 **Interactivity**   Annotating a photograph (int-3129)

2.7 Groundwater as a resource

2.7.1 What is groundwater?

An important part of the water cycle, groundwater is the water that is found under the Earth's surface. Many settlements — especially those in arid and semi-arid areas — rely on groundwater for their water supply.

When rain falls to the ground, some flows over the surface into waterways and some seeps into the ground. Any seeping water moves down through soil and rocks that are permeable; that is, they have pores that allow water to pass through them. Imagine pouring water into a jar of sand or pebbles; the water would settle into the spaces between the sand or stones.

Groundwater is water held within water-bearing rocks, or **aquifers**, in the ground. These work like sponges. They hold water in the tiny holes between the rock particles.

FIGURE 1 An artesian aquifer

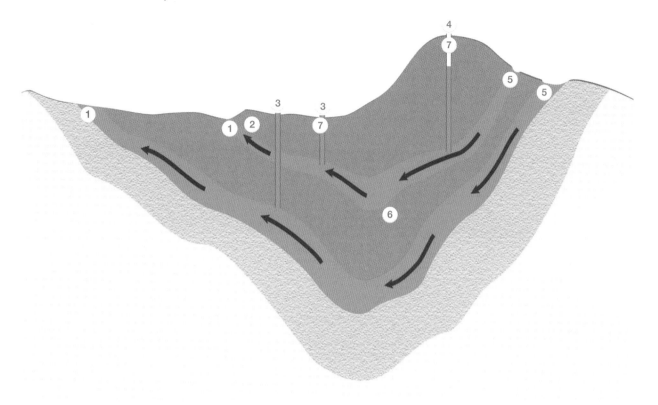

1   A spring is a place where water naturally seeps or gushes from the ground.

2   The watertable is a level under the land surface below which is saturated with water. This table rises and falls each year depending on surface rainfall and the amount of water taken from the ground.

3   Artesian bores are bores where the pressure of the water raises it above the land surface.

4   Subartesian bores are those where the water does not reach the surface and needs to be pumped out.

5   The aquifer is the saturated rock, usually made of sandstone, limestone or granite, where water is trapped between impermeable layers of rock.

6   Impermeable material rock layers do not allow water to flow through them.

7   Water flows freely from bores (wells) trapped in the aquifer.

2.7.2 Artesian water

An **artesian aquifer** occurs between impermeable rocks, and this creates great pressure. When a well is bored into an artesian aquifer, water often gushes out onto the surface. This flow will not stop unless the water pressure is reduced or the bore is capped (sealed).

Groundwater and surface water are interconnected — they depend on each other. Groundwater is only replenished when surface water seeps into aquifers. This is called groundwater recharge, and it is affected by whether there is a lot of rain or a drought is occurring.

FIGURE 2 The water in this mound spring in South Australia has taken over two million years to move to the surface from recharge areas in northern Queensland. It can take up to 1000 years to move about one metre.

2.7.3 Global groundwater resources

Approximately 2 per cent of the Earth's water occurs as groundwater, compared with 0.1 per cent as rivers and lakes and 94 per cent as oceans.

About 1.5 billion people in the world rely on groundwater for their survival. Some groundwater is fresh and can be used for drinking. Other groundwater can be brackish or even saltier than the sea.

FIGURE 3 The world's major groundwater basins

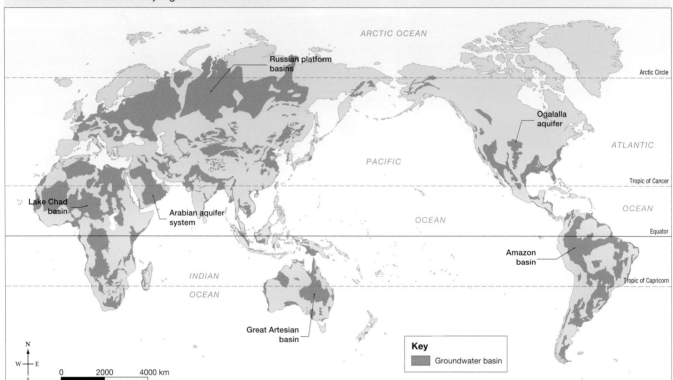

Source: BGR & UNESCO 2008: Groundwater Resources of the World 1 : 25 000 000. Hannover, Paris.

Groundwater is vital for drinking, irrigation and industry use. Some industries bottle and sell spring and mineral water, and make soft drinks and beer. Bore water is used to water suburban gardens and parks, golf courses and crops. Groundwater is also important to the natural environment in wetlands and in supporting unique plants and animals. Groundwater keeps many of our rivers flowing, even when there are long periods without rain.

Troubled waters

For many years now, more and more water has been taken out of the ground. People believed it was unlimited, but it is in danger of running out in some areas, owing to the large number of wells pumping water.

If people use more groundwater than is being recharged, aquifers may dry up. Groundwater is very slow-moving and can take many years to move into deep aquifers. For this reason, groundwater is a finite and non-renewable resource, and is often referred to as *fossil water*.

DISCUSS

Discuss the fairness of taking too much water from old aquifers. Should this type of water use be restricted? Who should make this decision? **[Critical and Creative Thinking Capability]**

2.7.4 How do Aboriginal peoples use groundwater?

Indigenous Australian peoples have lived in the Australian landscape since the beginning of the Dreaming, thousands of years by European estimates, and they have had the knowledge to survive many changes and challenges. In order to obtain water in the country's dry regions, particularly in Australia's deserts, they have needed to know where to find groundwater.

There are many groundwater sources throughout Australia that have long been used by Indigenous Australian peoples. One of these sources is **soaks**: groundwater that comes to the surface, often near rivers and dry creek beds, and which can be identified by certain types of vegetation. Another source is **mound springs**: mounds of built-up minerals and sediments brought up by water discharging from an aquifer.

Mound springs of the Oodnadatta Track

The Oodnadatta Track is located in the north-east of South Australia. The track follows the edge of the Great Artesian Basin and the south-western edge of Kati Thanda–Lake Eyre and, along its route, groundwater makes its way to the surface in several locations.

The Oodnadatta Track crosses the traditional lands of three Aboriginal nations. In the south, between Lake Torrens and Kati Thanda–Lake Eyre, are the Kuyani people; most of the west of Kati Thanda–Lake Eyre is the land of the Arabana people; and to the north is the land of the Arrernte people.

Many springs have cultural significance today for local Aboriginal peoples, whose ancestors relied on the springs as water sources and as sacred sites for important ceremonies. Knowledge of the springs in this region has been passed down over many generations through **Dreaming** stories.

FIGURE 4 Location of the Oodnadatta Track and Great Artesian Basin, one of the world's largest groundwater basins

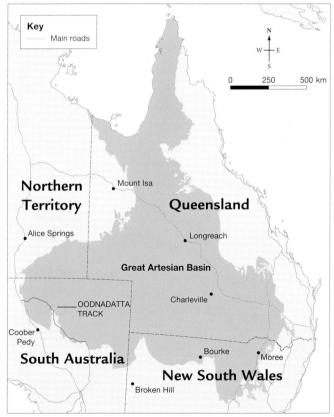

Source: Spatial Vision / Geoscience Australia

FIGURE 5 The Oodnadatta Track passes close to the Old Ghan, the Great Northern Railway.

This knowledge was also passed on to explorers and colonisers. John McDouall Stuart followed this track to complete the first crossing of Australia's interior from south to north in 1862; the overland telegraph

was constructed along its pathway; and the Great Northern Railway, which made the land of the Northern Territory accessible for European occupation, followed the same route.

Mound springs were very important for Indigenous Australian peoples. They could rely upon springs as reliable sources of water in a very harsh, dry environment. Old campsites and animal remains provide evidence that they remained there for varying time periods. However, because the plant and animal life around these regions is quite sparse, people had to move regularly and travel away from the springs when rainfall allowed that to occur.

Because the springs were strung out over hundreds of kilometres, they were also part of an important network of trading and communication routes across Australia. As Aboriginal peoples moved around the region, they traded goods and communicated with other Aboriginal nations. This interconnection allowed them to trade resources such as ochre, stone and wooden tools, bailer shells and pituri. Pituri is a spindly shrub used by Indigenous Australian peoples during ceremonies and to spike waterholes to catch animals for food.

2.7.5 CASE STUDY: Locating water

Stories help map the location of water

Indigenous Australian knowledge of the land and how to survive in it has been passed from generation to generation through Dreaming stories. During the dry seasons and periods of drought, Aboriginal peoples congregated at the mound springs. These springs were linked by Aboriginal songs and Dreaming stories, and are often connected to rain-making rituals.

FIGURE 6 Groundwater springs along the Oodnadatta Track

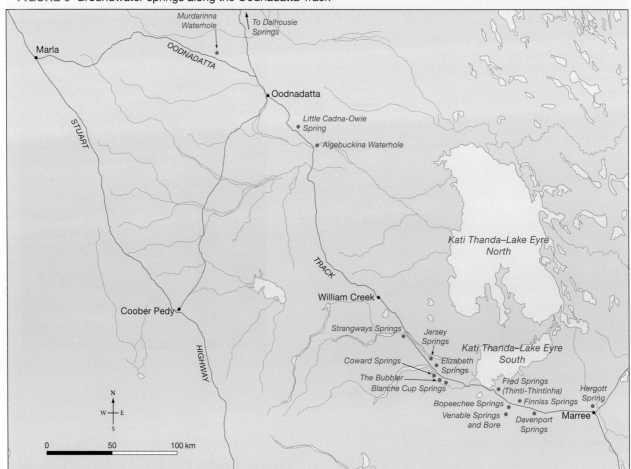

Source: Redrawn with permission from the SA Arid Lands Natural Resources Management Board / © Copyright Commonwealth of Australia Geoscience Australia 2006

Dreaming stories

1 Thutirla Pula (Two Boys Dreaming)

This is one of the most important stories of the Wangkangurru and other people of Central Australia. Thutirla Pula is how the spirits of the Dreaming first crossed the desert they call Munga-Thirri (land of sandhills). The story tells of two boys crossing the Simpson Desert, through Queensland and back to just north of Witjira (Dalhousie) in the Finke River area. The songline contains information on every waterhole or soak that was known in the Simpson Desert. Following this songline meant you could cross the Simpson Desert using available groundwater along the way, taking 600 kilometres off the usual journey south of the Simpson Desert to Kati Thanda–Lake Eyre, then back north along the Diamantina River.

2 Bidalinha (or the Bubbler)

The Kuyani ancestor Kakakutanha followed the trail of the rainbow serpent Kanmari to Bidalinha (or the Bubbler) where he killed it. He then threw away the snake's head, which is represented by Hamilton Hill, and cooked the body in a dirga, or oven, which is now Blanche Cup. Kakakutanha's wife, angry at missing out on the best meat from the snake, cursed her husband, and he went on to meet a gruesome death at Kudnangampa (Curdimurka). The bubbling water represents the movements of the dying serpent.

3 Thinti-Thintinha Spring (Fred Springs)

The willy wagtail (or thunti-thuntinha) danced his circular dance to create this spring and the surrounding soils, which are easily airborne in windy conditions. The moral to the story is that while it is easy to catch the skilful little willy wagtail, you must never do so because of the terrible dust storms that may follow.

4 Kewson Hill: The Camp of the Mankarra-kari — the Seven Sisters

The Seven Sisters came down here to dig for bush onions (yalkapakanha). As they peeled the onions, they tossed the skins to one side, creating the dark-coloured extinct mound spring on the south-west side of the track. The peeled bulbs created the light-coloured hill (yalka-parlumarna) to the north-east, also an extinct mound spring.

5 Dalhousie Springs

Dalhousie Springs is a popular oasis in the arid desert region of the northernmost part of South Australia.

FIGURE 7 The Old Bubbler on the Oodnadatta Track

FIGURE 8 Fed by the thermal waters of the Great Artesian Basin, the temperature of the water in Dalhousie Springs is between 34 and 38°C.

FIGURE 9 Cross-section of a typical mound spring

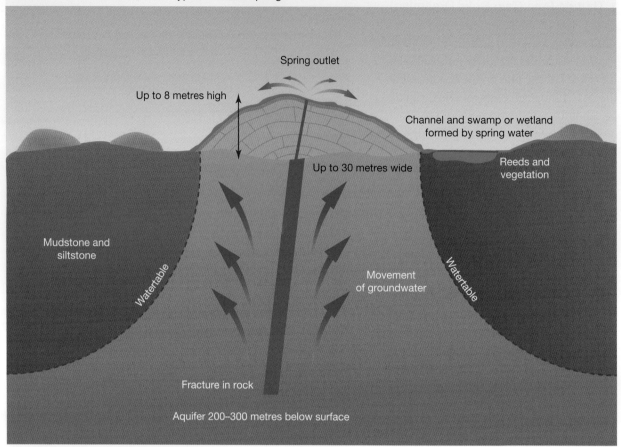

─Explore more with my**World**Atlas─

Deepen your understanding of this topic with related case studies and questions.
* Investigating Australian Curriculum topics > Year 7: Water in the world > **Salisbury Council — Aquifer storage, transfer and recovery**

 Resources

 Interactivities Water beneath us (int-3078)

Oodnadatta Track (int-3079)

2.7 INQUIRY ACTIVITIES

1. Use Google Earth and enter the search terms Oodnadatta or William Creek to locate the Oodnadatta Track. Describe the landscape you see. Why is finding groundwater so important in this *environment*?

   Describing and explaining

2. Investigate the consequences of countries over using fossil groundwater (groundwater regarded as a finite resource). Make a consequence chart and write 'Overuse of fossil groundwater' in the largest circle. Next write the consequences of this — you can add as many circles at the second and subsequent levels as you need. Continue to add as many consequences as you can. Now use a different coloured pen to suggest solutions to each of the consequences. Outline your final proposal to this issue once you have completed your consequence chart.

   Evaluating, predicting, proposing

2.7 EXERCISES

Geographical skills key: GS1 Remembering and understanding **GS2** Describing and explaining **GS3** Comparing and contrasting **GS4** Classifying, organising, constructing **GS5** Examining, analysing, interpreting **GS6** Evaluating, predicting, proposing

2.7 Exercise 1: Check your understanding

1. **GS1** What does the term 'permeable' mean?
2. **GS1** What is groundwater recharge?
3. **GS2** What is the difference between an aquifer and an artesian aquifer? Use a diagram to help you.
4. **GS2** Describe conditions that might result in a watertable rising or falling.
5. **GS2** Why were Dreaming stories important to Aboriginal peoples in the Oodnadatta region?
6. **GS2** How do Dreaming stories help map the groundwater in this region?

2.7 Exercise 2: Apply your understanding

1. **GS2** How do Dreaming stories help identify the cultural value placed on these water **environments**?
2. **GS2** Outline how groundwater and surface water are **interconnected**.
3. **GS2** Draw a diagram to show how surface water reaches the watertable to become groundwater.
4. **GS5** Refer to **FIGURE 3** to describe the location of the world's groundwater regions.
5. **GS6** Water is a renewable resource. Why is groundwater sometimes thought of as fossil water and as a non-renewable resource?

Try these questions in learnON for instant, corrective feedback. Go to www.jacplus.com.au.

2.8 Thinking Big research project: The Great Artesian Basin

SCENARIO

The federal government wants to supply Victoria and Western Australia with groundwater from the Great Artesian Basin (GAB). Your team of engineers has been commissioned by the Queensland, New South Wales, South Australian and Northern Territory governments to present a poster display to the federal Minister for Water convincing her that using the GAB to supply water to distant states is not a solution to the water crisis.

Select your learnON format to access:

- the full project scenario
- details of the project task
- resources to guide your project work
- an assessment rubric.

 Resources

projectsPLUS Thinking Big research project: The Great Artesian Basin (pro-0233)

2.9 Review

2.9.1 Key knowledge summary

Use this dot point summary to review the content covered in this topic.

2.9.2 Reflection

Reflect on your learning using the activities and resources provided.

 Resources

 eWorkbook   Reflection (doc-32131)

Crossword (doc-32132)

 Interactivity   Our precious environmental resources crossword (int-7699)

KEY TERMS

alluvial soil a fine-grained fertile soil brought down by a river and deposited on its bed, or on the floodplain or delta

aquifer a body of permeable rock below the Earth's surface that contains water, known as groundwater. Water can move along an aquifer.

atmosphere the layer of gases surrounding the Earth

artesian aquifer an aquifer confined between impermeable layers of rock. The water in it is under pressure and will flow upward through a well or bore.

biomass organic (once living) matter used as fuel

desalination a process that removes salt from sea water

the **Dreaming** in Aboriginal spirituality, the time when the Earth took on its present form and cycles of life and nature began; also known as the Dreamtime. It explains creation and the nature of the world, the place that every person has in that world and the importance of ritual and tradition. Dreaming Stories pass on important knowledge, laws and beliefs.

fossil fuels fuels that come from the breakdown of living materials, and which are formed in the ground over millions of years. Examples include coal, oil and natural gas.

groundwater a process in which water moves down from the Earth's surface into the groundwater

hydrologic cycle another term for the water cycle

intensive agriculture any method of farming that requires concentrated inputs of money and labour on relatively small areas of land; for example, battery hens and rice cultivation

manufacture to make products on a large scale

mound spring mound formation with water at its centre, which is formed by minerals and sediments brought up by water from artesian basins

riverine environment the environment around a river or river bank

soak place where groundwater moves up to the surface

subsistence farming a form of agriculture that provides food for the needs of only the farmer's family, leaving little or none to sell

turbine a machine for producing power, in which a wheel or rotor is made to revolve by a fast-moving flow of water, steam or air

uranium radioactive metal used as a fuel in nuclear reactors

3 Water use and management in the world

3.1 Overview

Our planet is covered with water, so why do we have to be so careful with it?

3.1.1 Introduction

Viewed from space, the Earth is a sphere of blue. Water covers most of our planet. We depend on water for life; in fact, no life is possible without it. Water is a precious and finite resource, yet most of the Earth's water is too salty for humans, animals or plants to use. The amount of available fresh water on Earth needs to be shared among an ever-growing global population. Access to water is a basic human right. It is a resource that must be used carefully so that current and future populations can have adequate supplies. The image shows what all of Earth's water would look like if it was contained in a sphere, in comparison with the size of the Earth. The blue sphere representing all of Earth's water has a diameter of 1385 kilometres.

on Resources

☑ **eWorkbook**   Customisable worksheets for this topic

🎞 **Video eLesson**   A world of water (eles-1616)

🔗 **Weblink**   Introduction to water

LEARNING SEQUENCE

3.1 Overview
3.2 Water in the world
3.3 Australia's climate and how it affects water availability
3.4 **SkillBuilder:** How to read a map
3.5 How do people use water?
3.6 **SkillBuilder:** Drawing a line graph   online only
3.7 Does everyone have enough water?
3.8 Virtual water
3.9 How can water be managed?
3.10 **Thinking Big research project:** Desalination plant advertising   online only
3.11 **Review**   online only

To access a pre-test and starter questions and receive immediate, **corrective feedback** and **sample responses** to every question, select your learnON format at www.jacplus.com.au.

3.2 Water in the world

3.2.1 The world's water

Water is vital to our survival and essential to most human activities. Although Earth seems blue in space, not much of the water we see is available for use. And of the useable fresh water that can be seen, access to it is unequal across the globe.

Water covers about 75 per cent of the Earth's surface. Yet, as **FIGURE 1** shows, almost all this water (97.5 per cent) is salt water and not available for human consumption. Only 2.5 per cent of the world's water is fresh, but most of this is also unavailable for use by people. More than two-thirds (69.5 per cent) of this fresh water is locked up in glaciers, snow, ice and permafrost. Of the remaining amount, 30.1 per cent is found in groundwater. Only 0.4 per cent is left — found in rivers, lakes, wetlands and soil as well as in the bodies of animals and plants.

FIGURE 1 The distribution of water on Earth

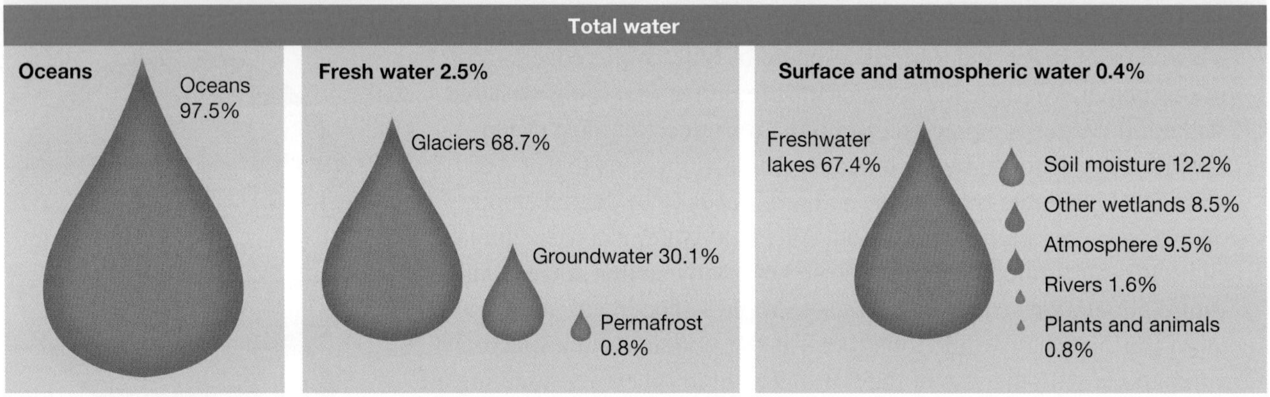

Total water
Oceans
Oceans 97.5%
Fresh water 2.5%
Glaciers 68.7%
Groundwater 30.1%
Permafrost 0.8%
Surface and atmospheric water 0.4%
Freshwater lakes 67.4%
Soil moisture 12.2%
Other wetlands 8.5%
Atmosphere 9.5%
Rivers 1.6%
Plants and animals 0.8%

FIGURE 2 The distribution of global rainfall

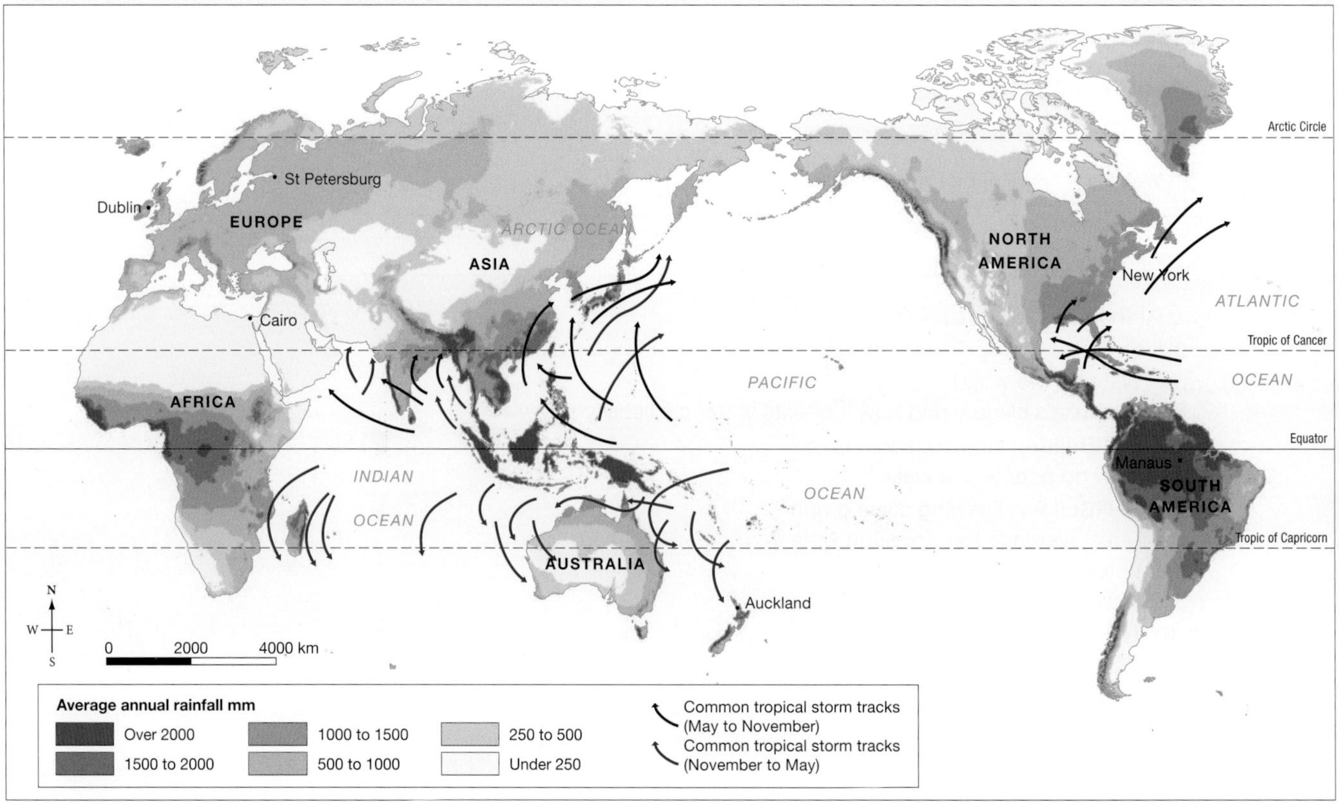

Arctic Circle
Tropic of Cancer
Equator
Tropic of Capricorn
St Petersburg
Dublin
EUROPE
ASIA
ARCTIC OCEAN
Cairo
AFRICA
INDIAN
OCEAN
AUSTRALIA
Auckland
PACIFIC
OCEAN
NORTH
AMERICA
New York
ATLANTIC
OCEAN
Manaus
SOUTH
AMERICA
N
W
E
S
0
2000
4000 km
Average annual rainfall mm
Over 2000
1500 to 2000
1000 to 1500
500 to 1000
250 to 500
Under 250
Common tropical storm tracks (May to November)
Common tropical storm tracks (November to May)

Source: WorldClim

Global rainfall

The Earth's water is constantly moving. Rainfall patterns show which world regions receive more rain than others. The amount of rainfall, or **precipitation**, is related to the amount of water available for people to use. **FIGURE 2** shows the distribution of global rainfall, and comparisons can be made between Australia and other continents and regions.

Green and blue water

The key to our survival is being able to use the water that falls on land and into rivers and streams. Water is sometimes categorised as either **blue water** or **green water**.

Green water is the water that does not run into streams or recharge groundwater but is stored in the soil or stays on top of the soil or vegetation. This water eventually evaporates or transpires through plants. Green water is used by crops, forests, grasslands and savannas.

The amount of blue and green water available changes throughout the year, from year to year, and according to changes in the environment.

3.2.2 Climate change and impact on rainfall and run-off

The majority of climate scientists believe that **climate change** will have an impact on rainfall patterns and **run-off**. Climate models have shown that areas in the northern latitudes are likely to experience more rain, and areas closer to the equator and mid-latitudes will receive less rain. Some regions will experience droughts, while others will experience high rainfall and even flooding.

Already, in the last 100 years, global rainfall patterns have changed. In some areas such as North America, South America, northern Europe, and northern and central Asia, rainfall has increased significantly. In other areas such as the Sahel, the Mediterranean, southern Africa, and parts of Asia, rainfall has decreased.

FIGURE 3 Predicted change in annual run-off due to climate change, 2084

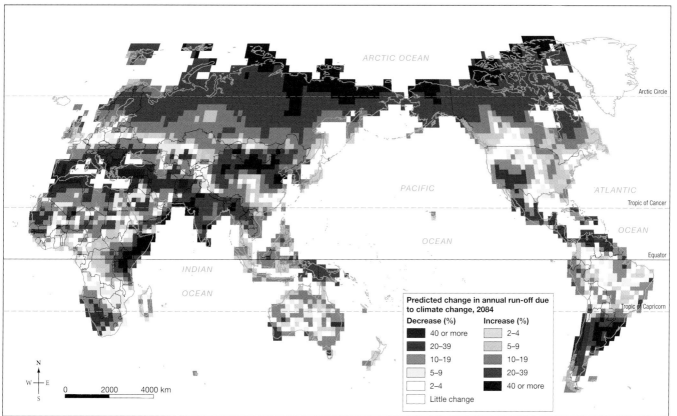

Source: Geophysical Fluid Dynamics Labratory, National Oceanic and Atmospheric Administration

3.2.3 Wonderful water

Although only 0.4 per cent of Earth's water is available for use by people, plants and animals, this amounts to over 117 trillion litres — a scale too big for us to really appreciate. There are some other amazing water facts that show how much water there is, how water connects places and how water is used.

How much water is there in the atmosphere?

If it was possible to rain across the whole planet at one time, there is enough water in the atmosphere at any time to produce about 2.5 centimetres (25 millimetres) of rain over the whole surface of the Earth.

FIGURE 4 A lot of water is held in the atmosphere.

The world's largest swimming pool

The artificial lagoon in Citystars Resort, Sharm el-Sheikh, Egypt, covers an area of 9.68 hectares. It is a full kilometre longer than the previous record holder in San Alfonso del Mar resort, Algarrobo, Chile, which measures 1013 metres in length and holds 250 000 cubic metres of water (see **FIGURE 5**).

The world's largest freshwater river

The Amazon River in South America is the second longest river in the world and is by far the largest by water flow. It has the largest **drainage basin** in the world and approximately one-fifth of the world's total river flow.

The **discharge** from it is about 7000 cubic kilometres every year, or 219 million litres of water every second. In contrast, Australia's largest river in terms of water flow, the Mitchell River, discharges an average of only 12 cubic kilometres each year.

The waterfall with the most water

The flow rate at Niagara Falls' main falls has been measured at around 2.6 million litres per second — enough water to fill an Olympic swimming pool in about one second.

FIGURE 5 Gargantuan pools are on the rise in luxury resorts around the world.

FIGURE 6 The river with the world's greatest water flow is the Amazon River.

FIGURE 7 Niagara Falls

The place on Earth with the most rainy days

The wettest place in the world (based on the average number of rainy days received each year) is Mount Wai'ale'ale in Hawaii. The summit is 1569 metres above sea level and receives over 350 days of rain each year. Mount Wai'ale'ale records an average of up to 13 000 millimetres of rain per year. In some years, rain has been known to fall for 360 days per year!

The wettest place in the world

The wettest place in the world (based on the yearly average total) is Mawsynram, India, which receives an average of 11 870 millimetres (nearly 12 metres) of rain each year. It has a subtropical highland climate with a long monsoon season.

FIGURE 8 Mount Wai'ale'ale averages over 350 rainy days each year.

FIGURE 9 Women in the potato fields of Mawsynram wear rain protection made of leaves and cane.

Where is the highest number of desalination plants in the world?

Places with little water often build desalination plants to convert sea water to fresh water. The Persian Gulf has the highest number of desalination plants in the world, with Saudi Arabia the world's largest producer.

The biggest dam

The Three Gorges Dam in China is the world's largest dam and hydro-electricity plant.

FIGURE 10 Al-Jubail, the largest desalination plant in the world, is located on the Persian Gulf in Saudi Arabia.

FIGURE 11 The Three Gorges Dam produces the most hydro-electricity in the world.

on Resources

Weblinks   Planet Earth, fresh water
              Iguazu Falls
Google Earth Al-Jubail
              Mawsynram
              Mt Wai'ale'ale
              Niagara Falls
              Three Gorges Dam

3.2 INQUIRY ACTIVITIES

1. Work in groups of three to list what might happen to people and the ***environment*** in regions that:
   (a) will receive more rainfall than they do now
   (b) will receive less rainfall than they do now.
   Complete a consequence chart for each ***change***.   **Evaluating, predicting, proposing**

2. Conduct some research to find the location and name of the longest river in the world and the river with the second-highest discharge. **Examining, analysing, interpreting**
3. Use a blank outline world map to locate the *places* mentioned in this **FIGURES 5**, **6** and **7**. Annotate your map with the water facts for which each *place* is famous. **Classifying, organising, constructing**
4. Conduct some research to find the following information: the *place* in the world where the most rain fell in a single 24-hour period; the *place* in Australia that has the record for the wettest year; the wettest town in Australia. Compare each of these with where you live. **Comparing and contrasting**
5. Watch the first half of the video clip on the **Iguazu Falls** weblink in the Resources tab.
   (a) Where are the Iguazu Falls located?
   (b) How wide are these falls?
   (c) How much water flows over the falls each second when they are in flood?
   (d) What is the size or *scale* of the flooded area of the Parana River?
   Remembering and understanding

3.2 EXERCISES

Geographical skills key: GS1 Remembering and understanding **GS2** Describing and explaining **GS3** Comparing and contrasting **GS4** Classifying, organising, constructing **GS5** Examining, analysing, interpreting **GS6** Evaluating, predicting, proposing

3.2 Exercise 1: Check your understanding

1. **GS1** What percentage of the world's water is:
   (a) salty
   (b) available for use by people?
2. **GS1** How will climate change affect rainfall patterns?
3. **GS3** Use the text to outline the difference between blue and green water. List two things that might *change* the amount of blue and green water available.
4. **GS2** If farmers use irrigation, what type of water would they rely on? What about farmers who do not have access to irrigation?
5. **GS1** How many Olympic-sized swimming pools would fit into the world's largest swimming pool?

3.2 Exercise 2: Apply your understanding

1. **GS5** Write a statement about the *interconnection* between high rainfall and location at the equator and mid-latitudes. Name two *places* that do not fit this pattern.
2. **GS5** Study **FIGURES 2** and **3** to answer the following questions.
   (a) Describe how much rain falls in North Africa and West Asia (the Middle East). How does this compare with Australia?
   (b) What is predicted to happen to annual run-off in these regions as a result of climate change? What impact might this have on people and the *environment*?
3. **GS5** Study **FIGURE 3** and an atlas.
   (a) Name three *places* that are predicted to receive more run-off due to climate change.
   (b) Name three *places* that are predicted to receive less run-off due to climate change.
   (c) Compare these six *places* with the global rainfall map, **FIGURE 2**. Which of the following statements is true?
      • Most *places* with very low rainfall have lower run-off.
      • All *places* with very high rainfall experience increased run-off.
      • The *places* with the greatest *change* in run-off will be northern Russia and northern Canada.
      Rewrite any false statements to make them true.
4. **GS1** What percentage of the world's freshwater is locked up in glaciers?
5. **GS1** Of the 0.4 per cent of water available to plants, animals and people, what percentage is in rivers?

Try these questions in learnON for instant, corrective feedback. Go to www.jacplus.com.au.

3.3 Australia's climate and how it affects water availability

3.3.1 Dry, variable and evaporated

Australia is the driest inhabited continent (only Antarctica is drier), and there is very little fresh water available for our use. Rain falls unevenly across the country and from season to season.

The driest part of Australia is around the Lake Eyre Basin, and the wettest locations are places in north-east Queensland and western Tasmania.

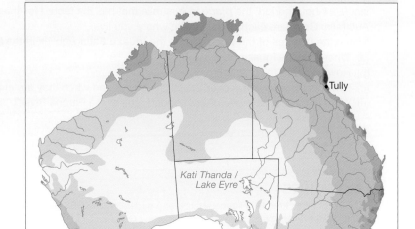

FIGURE 1 Australia's average annual rainfall

Source: Bureau of Meteorology 2003, on the Australian Water Map, Earth Systems Pty Ltd

3.3.2 Variability

Rainfall variability is the way rainfall totals in a given area vary from year to year. For example, if an area has low rainfall variability, it means rainfall will tend to be fairly consistent from one year to the next. Many coastal areas show this kind of rainfall pattern. In contrast, high rainfall variability means rainfall is likely to be irregular from one year to the next; there may be heavy rainfall in some years and little or no rainfall in others. Desert areas in central Australia tend to have low rainfall and high rainfall variability.

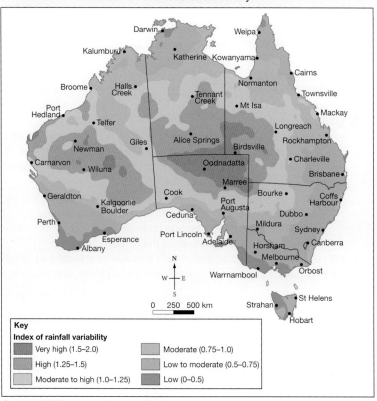

FIGURE 2 Australia's annual rainfall variability

Source: MAPgraphics Pty Ltd, Brisbane

FIGURE 3 Global rainfall variability

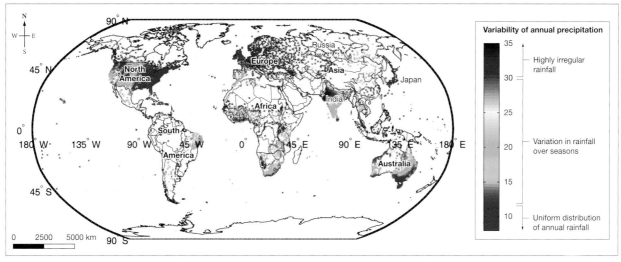

Source: Fatichi (2012)

3.3.3 Evaporation

Another problem for Australia is that most of its rainfall does not end up in rivers; much of it evaporates. Of all the water carried by the world's rivers, Australian rivers contain only 1 per cent of that total — even though Australia has 5 per cent of the world's land area. On average, only 10 per cent of our rainfall runs off into rivers and streams or is stored as groundwater. This figure drops to 3 per cent in dry areas and rises to 24 per cent in wetter places. The rest evaporates, is used by plants, or is stored in lakes, wetlands or underground storages. Areas in central Australia are very dry and, as a result, have high **evaporation** rates.

Relative humidity is a measure of the air's moisture content expressed as a percentage of the maximum moisture the air can contain at a certain temperature. Warm air can contain more moisture than cool air.

Relative humidity does not measure the exact amount of moisture in the air because that depends on air temperature. For example, if Brisbane has a day of 30 °C and Melbourne has a day of 15 °C, and the relative humidity in both places is 60 per cent, there will be much more moisture in the air in Brisbane than in Melbourne.

FIGURE 4 Average annual evaporation, Australia

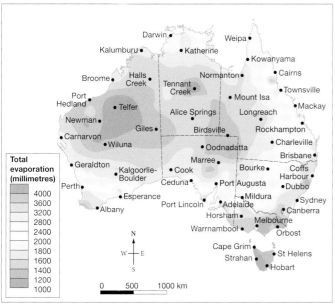

Source: Bureau of Meteorology

FIGURE 5 Average relative humidity across Australia

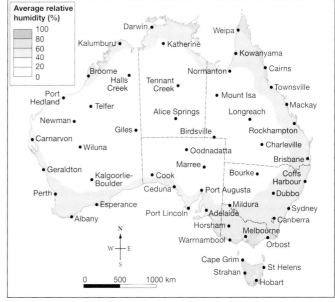

Source: Bureau of Meteorology

Relative humidity tends to be higher in coastal regions, as is rainfall, because areas with a lot of surface water have high evaporation. It is also higher in the parts of Australia that have very high rainfall, such as north Queensland and western Tasmania.

on Resources

Interactivity Hot and dry (int-3081)

Explore more with myWorldAtlas

Deepen your understanding of this topic with related case studies and questions.
- Australia > **Australia: weather and climate**

3.3 INQUIRY ACTIVITIES

1. Find the **place** where you live on the map of Australia. Study the four maps in this subtopic and complete a table like the one below. Compare where you live with another **place** in your state or territory and a **place** a long way from where you live.

Average rainfall	Rainfall variability	Average evaporation	Relative humidity
Where I live: _____ _____			
Another place in my state/territory: _____ _____			
A place far from where I live: _____ _____			

Comparing and contrasting

2. Australia has high evaporation rates and high rainfall variability. List all the ways that this **environment** makes water delivery to people a challenge. **Evaluating, predicting, proposing**
3. Find out the average rainfall for Kati Thanda–Lake Eyre and the wettest locations and heaviest rainfalls in north-east Queensland and western Tasmania. Record the rainfall variability, evaporation and relative humidity. What differences are there between these locations? **Comparing and contrasting**

3.3 EXERCISES

Geographical skills key: GS1 Remembering and understanding **GS2** Describing and explaining **GS3** Comparing and contrasting **GS4** Classifying, organising, constructing **GS5** Examining, analysing, interpreting **GS6** Evaluating, predicting, proposing

3.3 Exercise 1: Check your understanding
1. **GS2** What is rainfall variability?
2. **GS1** Which two regions receive the most rainfall in Australia?
3. **GS1** Which region has the most variable rainfall?
4. **GS2** What is relative humidity?
5. **GS2** Why does Australia have high evaporation rates?

3.3 Exercise 2: Apply your understanding
1. **GS5** Study **FIGURE 3**.
   (a) Name three countries or regions that have the highest rainfall irregularity (variability) in the world.
   (b) Name three regions that have the most uniform or reliable rainfall.
2. **GS5** Study the rainfall, humidity and evaporation maps (**FIGURES 1–4**). Fill in the missing word in the following statements in order to describe the **interconnections** between these features of our climate.
   Areas with low rainfall and low humidity tend to have a _____ evaporation rate.
   Areas with high rainfall and high humidity tend to have a _____ evaporation rate.

3. **GS5** Use **FIGURES 1**, **4** and **5** to record the following statistics for Tennant Creek in the NT.
   (a) Total evaporation
   (b) Average relative humidity
   (c) Average rainfall
4. **GS5** Use **FIGURES 1**, **4** and **5** to record the following statistics for Strahan in Tasmania.
   (a) Total evaporation
   (b) Average relative humidity
   (c) Average rainfall
5. **GS2** Where is the driest part of Australia located?

Try these questions in learnON for instant, corrective feedback. Go to www.jacplus.com.au.

3.4 SkillBuilder: How to read a map

What are maps and why are they useful?
Maps represent parts of the world as if you were looking down from above. Cartographers use colours and symbols on the map to show how features such as roads, rivers and towns are organised in a spatial way. Maps are useful to show features so that we have a deeper understanding of places.

Select your learnON format to access:
- an overview of the skill and its application in Geography (Tell me)
- a video and a step-by-step process to explain the skill (Show me)
- an activity and interactivity for you to practise the skill (Let me do it)
- questions to consolidate your understanding of the skill.

BOLTSS

B **Border** — a box around the map to clearly show its extent
O **Orientation** — a compass direction
L **Legend** — a key to what the symbols and colours on the map stand for
T **Title** — a clear indication of what the map is about or its theme
S **Scale** — indicates distances on the map compared with the actual area being shown
S **Source** — where possible, the information used to make the map should be sourced

on Resources

Video eLesson How to read a map (eles-1634)
Interactivity How to read a map (int-3130)

3.5 How do people use water?

3.5.1 What is water used for?
There are three main ways that all people use water: growing food, producing goods and electricity, and using it in the home. The amount of water consumed for each of these uses differs from one place to another. The problem remains that while the total amount of fresh water is fixed, the amount used per person is increasing.

It is interesting to look at water consumption on a global scale. With the global average at 1240 cubic metres per person, per year, some countries consume more water than others. Examples of countries that consume nearly twice as much as the global average are the United States and Thailand. Some countries that consume the least amount of water per person are Peru, Somalia and China.

FIGURE 1 shows that most of the world's water is used in agriculture, to grow food for the world's increasing population. This is especially the case in the drier parts of the world where there is not enough rainfall to grow crops or grass for animals. There is a strong interconnection between the amount of rainfall in a region and the amount of water used in agriculture.

It is interesting to see how this pattern varies in different countries. In some countries, the water used in agriculture and industry is greater than the amount of water used in homes for domestic use. In other places, people consume more water for domestic use than for either agriculture or industry.

FIGURE 1 Countries in the world differ in their use of water.

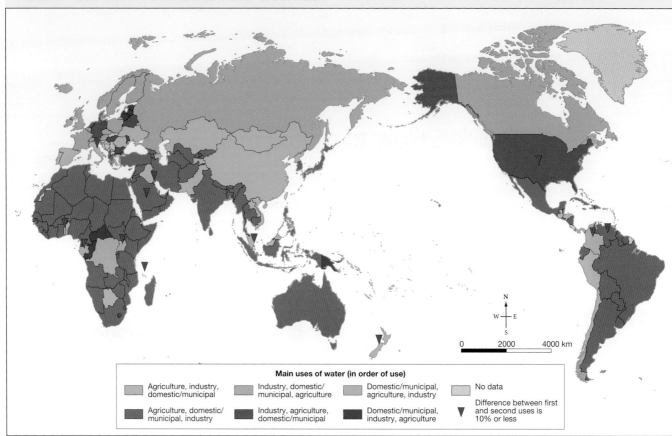

Source: Mekonnen, M.M. and Hoekstra, A.Y. 2011, 'National water footprint accounts: the green, blue and grey water footprint of production and consumption', *Value of Water Research Report Series No. 50,* UNESCO-IHE, Delft, the Netherlands.

FIGURE 2 Regional use of water for different purposes

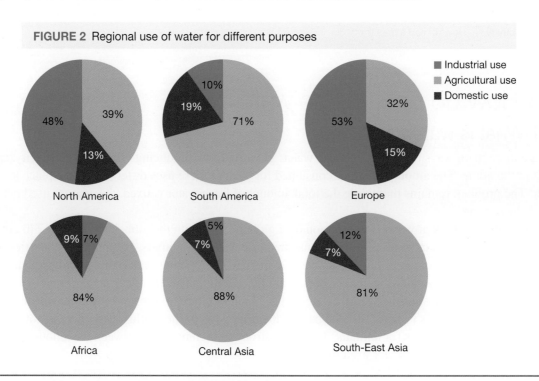

┌─ Explore more with my**World**Atlas ─────────────────────────

Deepen your understanding of this topic with related case studies and questions.
• Investigate additional topics > Managing water resources > **How is water used in Australia?**
• Exploring places > Europe > **North-west Europe**

3.5.2 How is water used in Australia?

Agriculture is an important industry in Australia, and it is our thirstiest industry. It produces most of our food requirements and contributes enormously to Australia's export earnings.

Around 70 per cent of Australia's fresh water is used as irrigation for farming. Many crops are grown in dry areas where up to half the available water evaporates from the soil surface or seeps down too low into the ground for plant roots to reach it. Therefore, more water is applied than is actually needed by plants. In manufacturing industries, most water is used to produce food, beverages and paper.

TABLE 1 Fresh water use in Australia

Types of use of fresh water	%
Agriculture • (pasture 35%) • (crops 27%) • (rural and domestic stock 8%)	70
Urban	12
Horticulture	10
Industry	3
Mining	2
Services	2
Hydro-electricity	1

In many areas in Australia where rainfall is limited or highly seasonal, farmers irrigate their crops with water stored in dams, with groundwater or with water from major rivers. Irrigation is a very important use of water in Australia. Most large-scale farming could not provide food for Australia's population without using water from rivers, lakes, reservoirs and wells.

There is high demand for irrigation water during summer when river flows are low, and low demand for irrigation water during winter when river flows are high. This reverses the natural pattern of river flow.

FIGURE 3 Australia is one of the most irrigated countries in the world.

TABLE 2 Fresh water used to irrigate different crops in Australia

Crop type	Water (gigalitres)	%
Livestock, pasture, grains and other agriculture	8795	56
Cotton	1841	12
Rice	1643	11
Sugar	1236	8
Fruit	704	5
Grapes	649	4
Vegetables	635	4

Note: One gigalitre = 1 000 000 000 litres or one thousand million litres or 400 Olympic-sized swimming pools

3.5.3 How do indigenous peoples use water?

Water is very important to both indigenous and non-indigenous peoples across the world, and is used for many different purposes. Water has a spiritual importance to the different indigenous peoples around the world, and many groups have strong connections with the natural world.

The extract below from the Indigenous Peoples Kyoto Water Declaration shows the importance of water. This declaration was put together in 2003 at the Third World Water Forum held in Kyoto, Japan.

We, the indigenous peoples from all parts of the world … reaffirm our relationship to Mother Earth and responsibility to future generations to … speak for the protection of water. We were placed in a sacred manner on this earth, each in our own sacred and traditional lands and territories … to care for water.

Source: www.waterculture.org/KyotoDeclaration.html

3.5.4 Aboriginal and Torres Strait Islander peoples

Depending on where Aboriginal and Torres Strait Islander peoples live, they collect surface water from creeks, rivers and waterholes; from underground water supplies such as soaks and springs; or directly from plants, including tree roots.

Evidence collected from oral histories, Dreaming stories, rock art, artefacts, ceremonial body painting and historical records left by colonists, missionaries, surveyors, invaders and explorers shows that Aboriginal and Torres Strait Islander peoples managed their water carefully. They channelled and filtered their water, covering it to keep it clean and to stop it from evaporating. They also created wells and tunnel reservoirs.

Indigenous seasons are closely linked to water. Use the **Miriwoong seasonal calendar** weblink in the Resources tab to see an example of one Indigenous calendar where seasons, water and activities are closely linked.

3.5.5 Water and Indigenous culture

Our cultural values of water are part of our law, our traditional owner responsibilities, our history and our everyday lives. Our [Anmatyerr] law has always provided for the values we place on water. Australian law should respect Anmatyerr Law so we can share responsibility for looking after water.

Source: The Anmatyerr people, in Rea, Dr N. & Anmatyerr Water Project Team 2008, 'Provision for cultural values in water management: the Anmatyerr story', *Land & Water Australia Final Report*, p. vi.

The rainbow serpent is a key symbol of creation, but its journey from underground to the surface also represents groundwater rising to the top via springs. The creation of water sources and where to find them was often told in stories or through artwork. The Dreaming story 'How the water got to the plains' is one story that describes how billabongs appeared in the dry inland plains. Use the **How the water got to the plains** weblink in the Resources tab to hear the story told by Butchulla elder Olga Miller.

FIGURE 4 Water is a strong symbol in Indigenous art.

An Aboriginal water project

On 30 March 2008, the Victorian Government returned the heritage-listed Lake Condah in Victoria to the Gunditjmara traditional owners. Lake Condah is considered one of Australia's earliest and largest aquaculture ventures. Aquaculture is the growing and harvesting of animals and plants in a water environment. The Gunditjmara people want to preserve their culture while engaging in tourism, water restoration and sustainability projects. One example is the plan to restore the ancient stone aquaculture system at the lake for eel farming. On 6 July 2019, Budj Bim Cultural landscape which includes Budj Bim volcano and Tae Rak (Lake Condah) was declared a UNESCO World Heritage site. This listing means Gunditjmara achievements will be recognised on a global scale and protection of the site will be increased.

FIGURE 5 Location map of Lake Condah

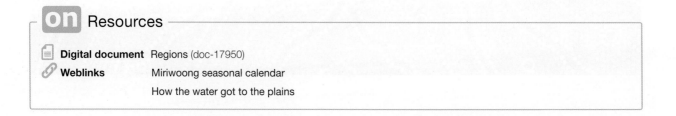

Casterton

Hamilton

Condah

Macarthur

Lake Condah

Heywood

Portland

Port Fairy

Key

—— Major roads

N
W   E
S

0   20   40 km

Source: Spatial Vision

on Resources

📄 **Digital document** Regions (doc-17950)

🔗 **Weblinks** Miriwoong seasonal calendar

How the water got to the plains

Explore more with myWorldAtlas

Deepen your understanding of this topic with related case studies and questions.
- Exploring places > Europe > **Mediterranean basin**
- Investigate additional topics > Managing water resources > **Three rivers in Africa**

3.5 INQUIRY ACTIVITIES

1. Research the Luangwa, Kafue and Zambezi rivers in Zambia, Africa. Use an atlas to trace the flow of each of the rivers and list some of the different water uses.   **Examining, analysing, interpreting**

2. Look at the Miriwoong seasonal calendar interactivity.
   (a) Which Australian region does this calendar represent?
   (b) How are the seasons divided? How does this compare with a European calendar?
   (c) How are seasons and water linked in this calendar?
   (d) How can all Australians benefit from Aboriginal and Torres Strait Islander peoples' knowledge and practices on seasons and water use in a regional context?

   Examining, analysing, interpreting

3. Use Google Earth to locate the Wabma Kadarbu Mound Springs Conservation Park. Place a pin on this location. Now zoom in and out to help you complete the following questions.
   (a) Where is this park located in South Australia? Where is this *place* in relation to where you live? Use distance and direction in your answer.
   (b) What is the name of the nearest road?
   (c) Describe the surrounding area.
   (d) Why would these springs be so important to Indigenous Australian peoples and European colonists?
   (e) Do some research to find out why these springs are protected today.
   (f) Use Google Maps to annotate this area with your findings. Include photos that you find using the internet.

   Examining, analysing, interpreting

4. Use the **Miriwoong seasonal calendar** weblink in the Resources tab to learn more about water seasons and food. How is water closely associated with Aboriginal activities in this region? **Describing and explaining**

3.5 EXERCISES

Geographical skills key: GS1 Remembering and understanding **GS2** Describing and explaining **GS3** Comparing and contrasting **GS4** Classifying, organising, constructing **GS5** Examining, analysing, interpreting **GS6** Evaluating, predicting, proposing

3.5 Exercise 1: Check your understanding

1. **GS1** What is most of the world's water used for?
2. **GS1** Consider water use around the world.
   (a) Which regions of the world use the majority of their water in agriculture?
   (b) Which countries use water mainly for industrial purposes?
   (c) Why might some countries use more water in industry than in agriculture or domestic use?
3. **GS1** From what sources did Indigenous Australian peoples collect water?
4. **GS1** What is aquaculture?
5. **GS2** What does evidence show about Aboriginal and Torres Strait Islander peoples' management of water supplies?

3.5 Exercise 2: Apply your understanding

1. **GS5** Study the data in **TABLES 1** and **2** to describe how water is used in Australia. Which crops use the most water? Which use the least?
2. **GS1** Study **FIGURE 1** and decide which of the following statements are true and which are false.
   (a) Australia uses most water for agriculture, then industry, then domestic/municipal.
   (b) Countries in North Africa use most water for industry, then domestic/municipal, then agriculture. (Use the **Regions** resource in the Resources tab).
   (c) Belarus uses most water for industry, then agriculture, then domestic/municipal.
   (d) Colombia uses most water for agriculture, then domestic/municipal, then industry.
   (e) Belize uses most water for industry, then agriculture, then domestic/municipal.
   (f) Malaysia uses most water for industry, then agriculture, then domestic/municipal.
   Rewrite the statements that are false and make them true.
3. **GS5** Use the orientation symbol in **FIGURE 5** to write the direction that the town of Hamilton is from the town of Condah. ▶

3.6 SkillBuilder: Drawing a line graph

What is a line graph?

A line graph displays information as a series of points on a graph that are joined to form a line. Line graphs are very useful to show change over time. They can show a single set of data, or they can show multiple sets, which enables us to compare similarities and differences between two sets of data at a glance.

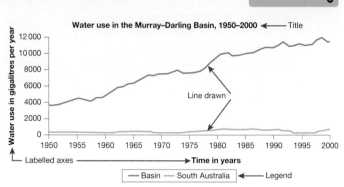

Select your learnON format to access:

- an overview of the skill and its application in Geography (Tell me)
- a video and a step-by-step process to explain the skill (Show me)
- an activity and interactivity for you to practise the skill (Let me do it)
- questions to consolidate your understanding of the skill.

Source: © Department of Environment, Water and Natural Resources, South Australia Government ← Source

on Resources

Video eLesson	Drawing a line graph (eles-1635)	
Interactivity	Drawing a line graph (int-3131)	

3.7 Does everyone have enough water?

3.7.1 The human right to water

The right to water is a human right that is protected by many international agreements, yet not everyone has access to this life-giving resource. Everyone has the right to enough safe, accessible and affordable water for all their needs. Water is more important to survival than food. In hot conditions, a person can survive up to three weeks without food but only two or three days without water.

People need access to **improved drinking water** yet, in 2018, water.org and World Vision reported that 844 million people were living without access to safe drinking water. Water is also needed to cook food, to bathe, to wash dishes and clothes, and to flush toilets. However, with the global population increasing and a fixed amount of water on Earth, some regions are suffering **water scarcity**. Water scarcity occurs when the demand for water is greater than the available supply.

Ideally, each individual needs one cubic metre (1000 litres) of drinking water per year, about 100 cubic metres for other personal needs, and 1000 cubic metres to grow all the food that he or she consumes. **Water stress** occurs when there is not enough water available for all demands. A country with less than 1000 cubic metres of renewable fresh water per capita (per person) is under water stress.

FIGURE 1 Distribution of global water scarcity

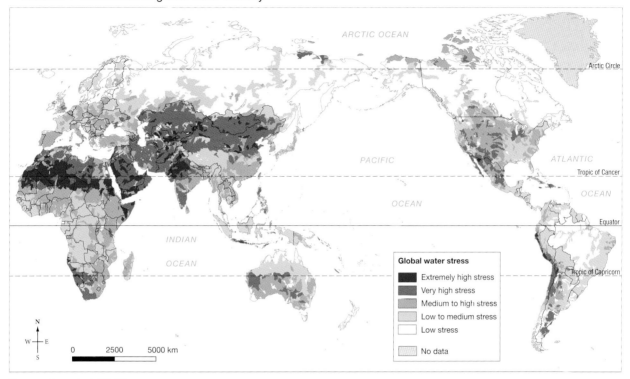

Global water stress

- Extremely high stress
- Very high stress
- Medium to high stress
- Low to medium stress
- Low stress
- No data

Source: Aqueduct (2019)

3.7.2 Access to water

A major reason that so many people lack access to safe water is the difference between where people live and where rain falls. Other reasons include water being used for agriculture and industry in regions where it is dry, and water being so polluted it cannot be used.

As climate change conditions take hold, the UN estimated in 2018 that 5 billion people could suffer water shortages by 2050. The problem of lack of water is often worse in rural areas, so many people move from the countryside into towns and cities, hoping for a better water supply. These people are sometimes called water refugees. However, the water in some cities is also inadequate because it is in short supply or is very polluted.

3.7.3 The water carriers

People who do not have water at home have to travel to get water. Water is very heavy and difficult to carry. The burden of this water fetching usually falls on women, who carry the heavy load on their head

FIGURE 2 Women often bear the burden of collecting water.

or back. For some people, the trip to a water supply and back can take hours each day — the average is 30 minutes to one hour. The average distance that women in Africa and Asia walk to collect water is six kilometres. The average weight they carry on their heads is about 20 kilograms — the usual weight of a suitcase taken on a flight. The World Health Organization estimates that over 40 billion work hours are lost each year in Africa alone just collecting drinking water.

FIGURE 3 Who carries and collects the water?

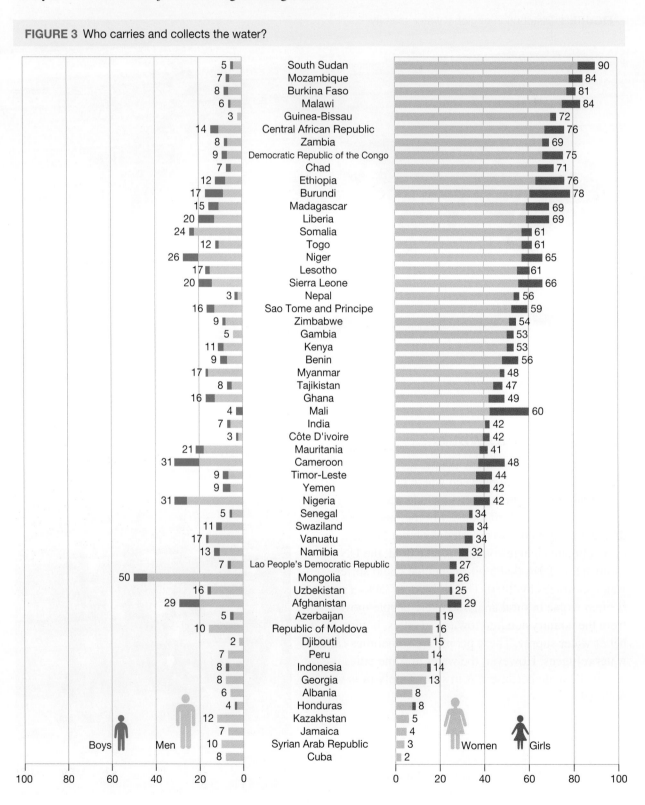

FIGURE 4 Drought conditions can reduce water quantity and quality.

DISCUSS
Conduct a class values continuum activity. Use pens or rulers end-to-end to create a line on the classroom floor. Standing on the left side of the line signifies you agree with a statement; on the right you disagree. If you stand a long way from the line, you strongly agree or disagree. If you stand on the line you don't know or are unsure. Be sure to be able to justify your stance.

Statement 1: It is too difficult to provide adequate clean water and sanitation to all the world's people.

When you have taken up your position, various class members give their reasons for their stance. Once these have been heard, read the statement again and see if you have changed your position on the continuum line. Repeat the activity with statement 2.

Statement 2: The Australian government should increase foreign aid so that more of the world's people can have access to clean water and adequate sanitation.

[Personal and Social Capability]

3.7.4 How does dirty water affect health?

Everyone in the world has the right to an adequate supply of water. The right to water is linked to many other rights, including the right to food and to health. In 2018 nearly 850 million people in the world had no access to clean water, and more than 2.3 billion people had no safe way of disposing of human waste. Lack of toilets means many people defecate in open spaces or near the same rivers from which they drink. It is estimated that 90 per cent of sewage in some countries ends up flowing straight into rivers and creeks.

This is an unacceptable situation. Dirty water and lack of proper hygiene kill around 315 000 children around the world every year, most of them younger than five. People who are sick are often unable to work properly, to look after their families or to attend school, keeping them trapped in the poverty cycle. The diseases that can be passed on to people as a result of contaminated water include diarrheal diseases such as cholera, typhoid and dysentery. Malaria, a disease transmitted by mosquitoes, kills about one million people every year.

3.7.5 Polluted rivers

Water quality can affect health in many ways. Rivers and streams act as drainage systems and, when it rains, water transports rubbish, chemicals and other waste into drains and, eventually, rivers.

Different pollutants — faeces (human and animal), food wastes, pesticides, chemicals and heavy metals — can come from industrial wastewater, domestic sewage, cars, gardens, farmland, mining sites and roads, and flow into waterways.

Some countries, cities and local areas are better than others at providing services and enforcing laws to prevent pollutants from entering water. Some of the worst polluted rivers and lakes in the world include rivers and aquifers in China (such as the Songhua River and the Yellow River), the Citarum River in Indonesia, the Yamuna and Ganges rivers in India, the Buriganga River in Bangladesh and the Marilao River in the Philippines.

FIGURE 5 The Citarum River in Indonesia is one of the most polluted rivers in the world.

3.7.6 How clean is your river?
Fieldwork: Investigating waterways

Some schools are located close to a waterway, even if it is a highly modified one like a concrete drain. Conducting fieldwork at a local waterway will help you to better understand national and global issues.

The aim is to investigate the physical properties of a river or creek at various points along its length, and to make observations of the water quality and any evidence of human impact. Does the quality of the water change between upstream and downstream sites? Are there human factors that can account for these changes? Differences may be more obvious if the waterway passes through a built-up area or a farm.

The following activities should be undertaken at each site and recorded on paper or directly onto a database on a laptop computer or other mobile device. Use a map and camera to record observations about the surroundings of each site, especially the amount of vegetation and possible human impact. Use GPS to record and map your location at each site.

Measuring river width

Stretch a tape measure 20 centimetres above the water from one bank to the other, measuring from where the dry bank meets the water. Take your reading directly above the tape at several locations and calculate an average.

Measuring the water depth

While the tape measure is stretched across the river or creek, use a metre rule to measure the depth. Record the depth every 50 centimetres (or 30 centimetres if the creek is small). Make sure the ruler only just touches the riverbed, and record each measurement as it is taken.

Temperature

Aquatic plants and animals have a particular temperature range in which they can survive. High water temperatures can result in reduced oxygen available for plants and animals. It is useful to compare temperature readings with **biodiversity** counts to investigate this relationship. Place the bulb of a thermometer in the water for five minutes and record the result.

pH

We use pH to measure the acidity or alkalinity of water on a scale of 1 to 10. Drinking water should have a pH reading of around 6. A reading either side of this may indicate that water is polluted. You can test pH by taking a sample of water and using pH paper or chemical reagents.

Turbidity

Water that lets little sunlight through is said to be turbid. Turbidity is the amount of suspended sediment in water — sediments such as clay, silt, industrial waste or sewage. A Secchi disc is used to measure turbidity. This can be made using an ice-cream lid, string, a weight and black paint.

Lower the disk into the water, making sure there are no waves or ripples, cloud or glare. Do not wear sunglasses when making the reading. When it is only just visible, record the depth in centimetres. The lower the number, the greater the turbidity.

Salinity

Salinity measures the amount of salt in the water. To measure salinity, you will need to use an electrical conductivity meter, or EC meter, which can be bought from science equipment suppliers.

Biodiversity

Biodiversity in water is studied by taking a small area of water and investigating the number and diversity of animal species. It is measured by ponding or water sampling. Choose a number of sites in the water and ensure that they contrast with each other: there should be clear, muddy, deep, shallow, moving and still sites, for example.

The materials needed for ponding include a fine net; a white plastic ice-cream container; a magnifying glass; a notebook and pencil for recording the number and variety of species; a camera for photographing specimens; and an identification chart or book.

Aesthetics — what the water looks like

Another measurement of water quality is to grade the appearance of the water. If an area of water is appealing to look at, it is considered aesthetically appealing. Observe aspects such as colour, odour, and the presence of algae, surface film or oil slicks. Ratings from 1 (excellent) to 5 (extremely poor) can be used and recorded. Photographs and field sketches of the sites are also useful.

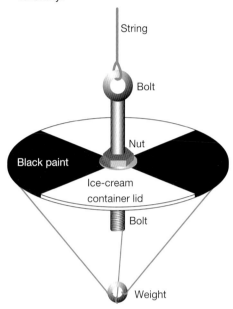

FIGURE 6 You can create a simple Secchi disc and use it to measure turbidity.

FIGURE 7 Students measuring river water quality

3.7.7 How can water-borne diseases be reduced?

People use different methods to treat the water they have collected. They can let it stand and settle, strain it through a cloth, filter it, add bleach or chlorine, or boil it. Some people do not treat their water at all.

When there is barely enough water to drink or to cook with, it is difficult for people to set aside water for washing hands and cleaning clothes. However, hygiene and sanitation are very important for health.

There are many aid groups (such as Water.org, Clean Water Fund, Global Water Challenge and The Water Project) who work on projects to improve sanitation and access to clean water. Washing hands, building cheap and effective toilets, and teaching the community about good hygiene all help to reduce disease.

FIGURE 8 Collecting water that is unsafe to drink

3.7.8 Sustainable Development Goals

Developed by the United Nations, the Sustainable Development Goals (SDGs) came into force on 1 January 2016. They are goals that aim to end all forms of poverty, protect the planet and ensure that all people enjoy peace and prosperity, with targets to be reached by the end of 2030. The SDGs build on the success of the Millennium Development Goals (MDGs) that were adopted from 2000–2015.

Goal 6 of the SDGs is to 'Ensure access to water and sanitation for all'. From 1990 to 2015 — including the 15 years of the Millennium Development Goals — the percentage of people who had access to clean water increased from 76 to 91 per cent.

Some of the targets (by 2030) for Goal 6 are to:
- provide universal and equitable access to safe and affordable drinking water for all
- provide access to adequate and equitable sanitation and hygiene for all and end open defecation, paying special attention to the needs of women and girls and those in vulnerable situations
- improve water quality by reducing pollution, eliminating dumping and minimising release of hazardous chemicals and materials, halving the proportion of untreated wastewater and substantially increasing recycling and safe reuse globally
- substantially increase water-use efficiency across all sectors and ensure sustainable withdrawals and supply of freshwater to address water scarcity and substantially reduce the number of people suffering from water scarcity
- protect and restore water-related ecosystems, including mountains, forests, wetlands, rivers, aquifers and lakes (this goal is by 2020).

FIGURES 9 and 10 show access to improved water and sanitation throughout the world.

FIGURE 9 Percentage of global population with access to improved water facilities, 2015

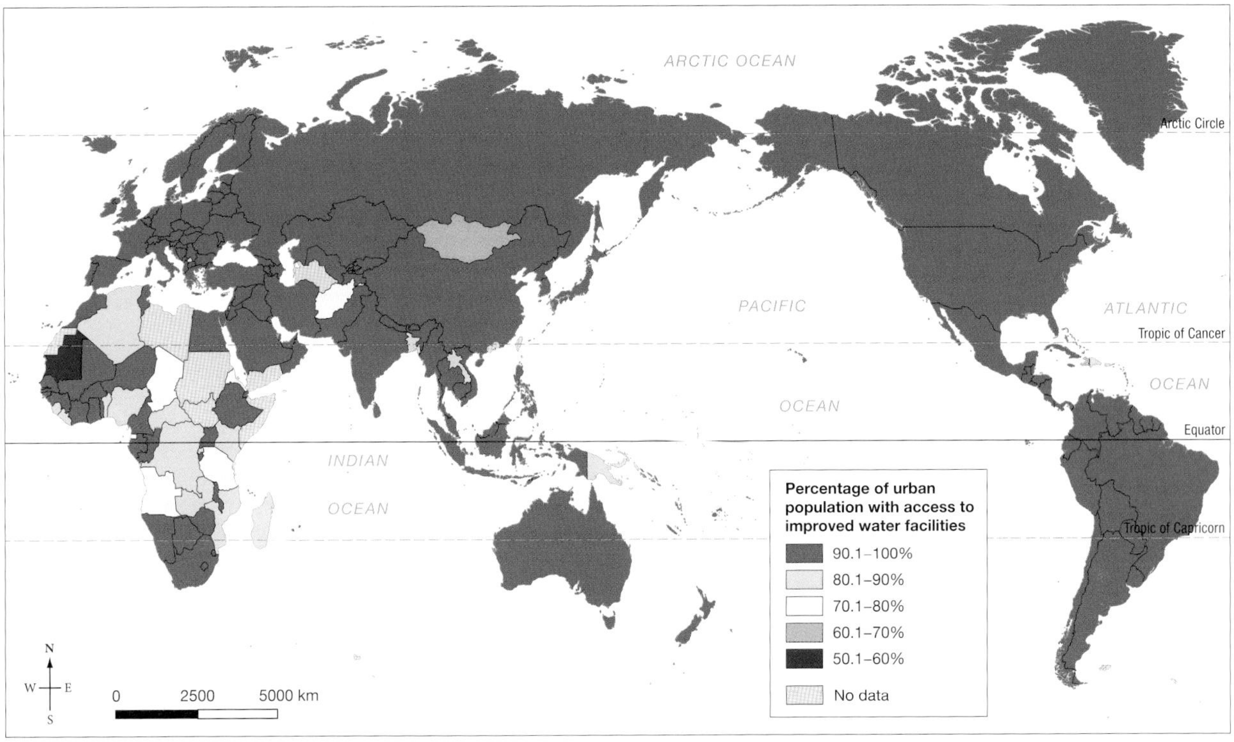

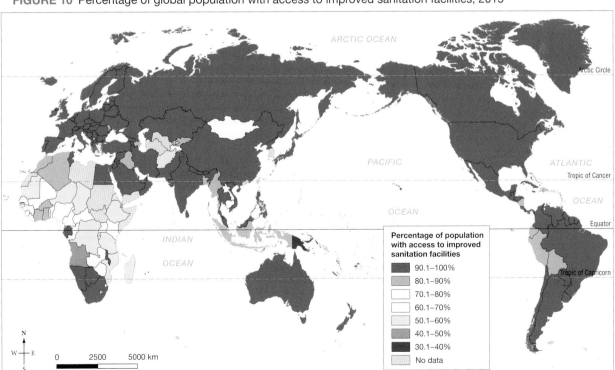

Source: World Bank (2015)

FIGURE 10 Percentage of global population with access to improved sanitation facilities, 2015

Source: World Bank (2015)

3.7.9 CASE STUDY: Community-led total sanitation (CLTS), Nigeria

From 2012 to 2016, Action Against Hunger worked with 138 communities in northern Nigeria using the community-led total sanitation (CLTS) methodology. According to the World Health Organization, in 2015 diarrheal diseases caused over 75 000 deaths in children aged 1–59 months. This is because access to improved sanitation in rural areas is limited. When CLTS leaders arrived in this area of northern Nigeria, there were very high rates of illness and death due to water-related diseases. The people in these villages used the bush near their houses as their toilet.

CLTS leaders developed a relationship with the villagers and taught them how faeces can enter their food and water, and make them sick. Once this was understood, the people wanted to change their practices so this would no longer happen. Action plans were drawn up and eventually the villagers created clean water points, built simple but effective toilets and were given lessons in sanitation.

The study has shown promising outcomes for these communities. Overall, 23 out of 46 communities (50 per cent) reported no open defecation among adults or children at home or away from home. This demonstrates that communities can make lasting behaviour changes. The study also showed that people face challenges such as collapsing toilets, habit and tradition, and financial difficulties, all of which need to be overcome.

FIGURE 11 Location of Nigeria in Africa

Source: Spatial Vision

Explore more with myWorldAtlas

Deepen your understanding of this topic with related case studies and questions.
- Investigating Australian Curriculum topics > Year 7: Water in the world > **The Dead Sea — overcoming water scarcity**
- Exploring places > Europe > **Russia and Eurasia**
- Exploring places > The world > **World: health**

---

on Resources

Digital document	Regions (doc-17950)
Weblinks	Burden of thirst
	Health is wealth
	UN SDG
	WHO/UNICEF
Google Earth	Citarum River

3.7 INQUIRY ACTIVITIES

1. Study **FIGURE 3**. Women are the main water carriers in *places* where there is water stress and scarcity. Discuss with another student how this would affect a woman's health, education, family life and food production. Draw a consequence map of your ideas. **Classifying, organising, constructing**
2. Geographers like to look at patterns over *space*. Find an atlas map showing population density and compare it with **FIGURE 1**. Refer to countries in North Africa and West Asia in your answers. (Refer to the **Regions** resource in the Resources tab.)
   (a) Name three coastal and two landlocked countries in these regions.
   (b) Conduct research to find out how many countries in these regions suffer deaths caused by poor sanitation and dirty water.

(c) Name the continents and regions that have the fewest deaths.

(d) Why do you think this *spatial* pattern exists? (*Hint:* Look at maps in your atlas that show wealth.)

Examining, analysing, interpreting

3. What would be the impact on the lives of women and children if there was a well with clean water in every village or town? Discuss in small groups the fairness or otherwise of this situation. How could this be improved for women and children? **Evaluating, predicting, proposing**

4. Work in groups of three or four. Use the data and facts in these pages to plan a day of promoting knowledge about this issue at your school. Use the links available at the **UN SDG** weblink in the Resources tab which includes information on programs. Make particular reference to north and sub-Saharan Africa, and find out what is being done by aid organisations to improve the situation in these regions. Plan a video presentation that is interesting and catchy and will help people understand the action needed to improve access to clean water and sanitation. Use video and video editing programs and internet research in your planning.

Classifying, organising, constructing

5. A number of aid agencies are working in countries and regions to improve access to sanitation and clean water. Choose one of those listed in this topic and find out more about what they are doing. How will their work make a difference to the living conditions of the people they are helping?

Evaluating, predicting, proposing

6. Use the **Burden of thirst** weblink in the Resources tab to watch a video about water scarcity in east Africa.

Examining, analysing, interpreting

7. Use the **Health is wealth** weblink in the Resources tab to learn about a community-led sanitation project.

Examining, analysing, interpreting

3.7 EXERCISES

Geographical skills key: GS1 Remembering and understanding **GS2** Describing and explaining **GS3** Comparing and contrasting **GS4** Classifying, organising, constructing **GS5** Examining, analysing, interpreting **GS6** Evaluating, predicting, proposing

3.7 Exercise 1: Check your understanding

1. **GS2** Why is access to water a human right?
2. **GS2** What is meant by water scarcity and water stress?
3. **GS3** What is the difference between improved and unimproved water supplies?
4. **GS2** Describe the impact on a country if it is under water stress or water scarcity.
5. **GS2** There is a difference in water quality across Australia. Is your water supply 'improved'? Explain.
6. **GS1** What might be meant by the term 'water refugee'?
7. **GS2** How do rivers become polluted?
8. **GS1** How many people in the world do not have access to clean water or sanitation? How do the SDGs aim to improve this situation?
9. **GS2** Describe Nigeria's location in Africa and in relation to other countries and *places*.
10. **GS1** What does CLTS stand for?

3.7 Exercise 2: Apply your understanding

1. **GS2**
   (a) Use an atlas to help you identify and list three countries that are experiencing extremely high water stress.
   (b) What is Australia's level of water stress compared to New Zealand?
   (c) Name the largest country in South America with low water stress.
2. **GS5** Use **FIGURE 1** to describe the water scarcity in north Africa.
3. **GS5** Study **FIGURE 3**.
   (a) Of the 55 countries listed, how many African countries are represented in this graph?
   (b) Which two countries have the highest representation of girls carrying water?
   (c) Which country is the only one where boys and men are the primary water carriers?
   (d) Which Asian country has the highest representation of girls and women carrying water?
4. **GS2** Explain what community-led total sanitation projects aim to do.
5. **GS6** What might happen to people's health in north and sub-Saharan Africa if access to water and sanitation is not improved?

Try these questions in learnON for instant, corrective feedback. Go to www.jacplus.com.au.

3.8 Virtual water

3.8.1 What is virtual water?

The water we consume is not just what we use in cooking, drinking, washing, flushing the toilet or in the garden. Water is used to manufacture everything we use: mobile phones, toys, cars and newspapers. This **virtual water** needs to be considered in our **water footprint**.

Virtual water is also known as embedded water, embodied water or hidden water. It includes all the water used to produce goods and services. Food production uses more water than any other production.

Hidden in a cup of coffee are 140 litres of water used to grow, produce, package and ship the beans. That is roughly the same amount of water used by an average person daily in Australia for drinking and household needs. There is a lot of water hidden in a hamburger too: 2400 litres. This includes the water needed to grow the feed for the cattle over a number of years, to grow wheat for the bread roll, to grow all the other ingredients in the hamburger, and to process all the food.

Virtual water varies from food to food. For example, it takes about 3400 litres of water to grow one kilogram of rice, whereas it takes 200 litres to grow one kilogram of cabbage. Regions that are water stressed and that export food and other products (such as Australia and some countries in Africa and Asia) are also effectively exporting their precious water in these goods.

A country that imports rice, rather than growing it locally, therefore saves 3400 litres of water for every kilogram it imports. Some countries (such as Japan) have very little land on which to grow food; other countries have very few cubic metres of renewable water per person. Singapore, for example, has only about 130 cubic metres per person. Both types of countries survive by virtual water imports: they import food rather than attempt to grow and produce all their food themselves. This means that small, wealthy countries can import food that needs a lot of water to produce, and export products that need little water to produce. This makes water available for other domestic purposes such as drinking and cooking.

TABLE 1 The water used to grow food and provide goods varies from product to product.

Food item	Unit	Global average water (litres)
Apple or pear	1 kg	700
Barley	1 kg	1300
Banana	1 kg	860
Beef	1 kg	15 500
Beer (from barley)	250 mL	75
Bread (from wheat)	1 kg	1300
Cabbage	1 kg	200
Cheese	1 kg	5000
Chicken	1 kg	3900
Chocolate	1 kg	24 000
Coconut	1 kg	2500
Coffee (roasted)	1 kg	21 000
Cotton shirt	1	2700
Cucumber or pumpkin	1 kg	240
Dates	1 kg	3000

(continued)

TABLE 1 The water used to grow food and provide goods varies from product to product. *(continued)*

Food item	Unit	Global average water (litres)
Eggs	1	200
Goat meat	1 kg	4000
Groundnuts (in shell)	1 kg	3100
Hamburger	1	2400
Lamb	1 kg	6100
Leather	1 kg	16 600
Lettuce	1 kg	130
Maize	1 kg	900
Mango	1 kg	1600
Millet	1 kg	5000
Milk	250 mL	250
Olives	1 kg	4400
Orange	1	50
Paper	1 A4 sheet	10
Peach or nectarine	1 kg	1200
Pork	1 kg	4800
Potato	1 kg	4800
Rice	1 kg	3400
Soybeans	1 kg	1800
Sugar (from sugar cane)	1 kg	1500
Tea	250 mL	30
Tomato	1 kg	180
Wheat	1 kg	1300
Wine	125 mL	120

The major exporters of virtual water are found in North and South America (the United States, Canada, Brazil and Argentina), south Asia (India, Pakistan, Indonesia, Thailand) and Australia. The major virtual water importers are north Africa and the Middle East, Mexico, Europe, Japan and South Korea.

DISCUSS

'In a dry country such as Australia, it makes more sense to import food and resources that require a lot of water to produce'. Relating to your understanding of virtual water, provide an argument for this viewpoint and an argument against this viewpoint. Ensure that your argument is supported with evidence and is logical.

[Critical and Creative Thinking Capability]

3.8.2 What is a water footprint?

The water footprint of an individual or country is the total volume of fresh water that is used to produce the goods and services consumed by the individual or country. It includes the use of:

- blue water (rivers, lakes, **aquifers**)
- green water (rainfall used for crop growth)
- grey water (water polluted after agricultural, industrial and household use).

FIGURE 1 Average global water footprints

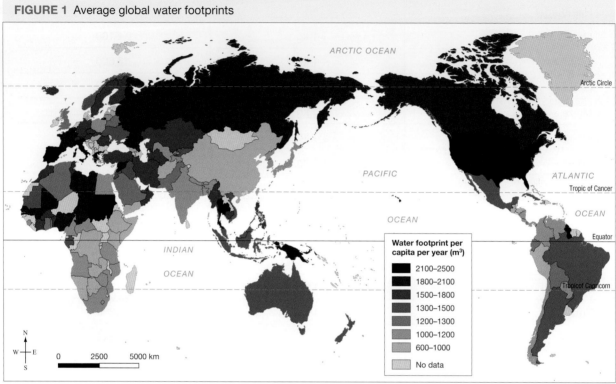

Water footprint per capita per year (m³)

- 2100–2500
- 1800–2100
- 1500–1800
- 1300–1500
- 1200–1300
- 1000–1200
- 600–1000
- No data

Source: World Water Exchange (2016)

Not all goods consumed in one particular country are produced in that country — some foods and products are imported. Therefore, the water footprint consists of two parts: use of domestic water resources and use of water outside the borders of the country.

In the United States, the average water footprint per year per capita is 2480 cubic metres per person per year, which is enough to fill an Olympic swimming pool. In China, the average water footprint is 1071 cubic metres per year. The figure for Australia is 1393 cubic metres per person per year

Japan, with a footprint of 1200 cubic metres per person per year, has about 65 per cent of its total water footprint outside its borders, meaning a lot of its water is imported in the form of consumer goods and food. On the other hand, only about 7 per cent of the Chinese water footprint falls outside China.

┌─Explore more with my**World**Atlas──────────────

Deepen your understanding of this topic with related case studies and questions.
- Investigate additional topics > Managing water resources > **Our water footprint**

┌─ **on** Resources ─────────────

📄 **Digital document**   Regions (doc-17950)
🔧 **Interactivity**   Unreal (int-3080)
🔗 **Weblink**   Just add water

1. Use a blank world map and choose two colours to show the main importers and exporters of virtual water. Describe the *spatial* patterns of the map you have drawn.
2. Conduct a debate on the following statement: 'People should eat less meat in order to consume less water.'
3. Use the **Just add water** weblink in the Resources tab to listen to an audio program about the water footprint in food production.
   (a) What is the relationship between water-stressed countries and food production?
   (b) Give an example where the water footprint figure is in conflict with the opinion of farmers.
   (c) Detail a product produced in Wodonga that includes both virtual and domestic water.

3.8 EXERCISES

Geographical skills key: GS1 Remembering and understanding **GS2** Describing and explaining **GS3** Comparing and contrasting **GS4** Classifying, organising, constructing **GS5** Examining, analysing, interpreting **GS6** Evaluating, predicting, proposing

3.8 Exercise 1: Check your understanding

1. **GS2** Explain the difference between virtual water and a water footprint.
2. **GS3** Outline the differences between blue, green and grey water.
3. **GS2** Refer to **FIGURE 1**. Describe the patterns you notice over *space* of countries with (i) very high and high water footprints and (ii) very low and low water footprints.
4. **GS2** Describe the water embedded (or hidden) in a cup of coffee.
5. **GS2** Describe the water embedded (or hidden) in a hamburger.

3.8 Exercise 2: Apply your understanding

1. **GS3** Study **TABLE 1**. Choose three meat, five grain, two dairy, two non-food, four fruit, four vegetable and two processed products from the list. Create a bar graph to show how much water is used to produce a vegetarian diet and a meat-based diet. Which diet uses more water?
2. **GS3** Compare the data for north Africa and west Asia (the Middle East) in **FIGURE 1** and the maps in subtopics 3.2, 3.5 and 3.7. Write three summary statements that describe the amount of rainfall, water use (including water footprints) and water availability for these two regions. How do these patterns compare with Australia?
3. **GS2** Describe how water footprints connect *places* often far away from each other.
4. **GS2** Why are countries that export goods and food said to be also exporting their water?
5. **GS2** Why is it that countries with little water and land are said to import water when they import food?

Try these questions in learnON for instant, corrective feedback. Go to www.jacplus.com.au.

3.9 How can water be managed?

3.9.1 Managing our water supply

More water cannot be created, but it can be managed better. With a growing global population, and the predicted changes due to climate change, the pressure on this finite resource requires a number of solutions.

Introducing effective water management can be a challenge at any scale, whether local, national or global. It needs the cooperation of all users, including farmers, industry, individuals, and upstream and downstream people in different countries or different states. With all the competing demands on water, management is often easier to approach at a local scale.

Because agriculture uses the greatest amount of water, it makes sense to make irrigation systems more efficient. The aim is to get more production for every drop of water used. Some irrigation systems waste up to 70 per cent of their water through leaks and evaporation, so changing the irrigation method can save water. Other management practices include recycling, using desalinated water and using stormwater.

3.9.2 Managing water across borders

About 260 drainage basins across the world are shared by two or more countries. Thirteen river basins are shared by five or more countries. Depending on their location in the catchment, some countries can suffer reduced access to water because of other countries' usage. This shows the *interconnection* between places — what happens in one place affects another. Diverting rivers, building dams, taking large amounts of water out for irrigation, and creating pollution can all lead to conflict between countries, states and political groups.

Country disputes have occurred in the Nile Basin in north Africa, along the Mekong River in Asia, the Jordan River Basin in west Asia (the Middle East) and along the Silala River in South America. This can also happen within a country, which has happened with the Murray–Darling Basin in Australia, across four states and one territory.

Some countries sign international agreements or treaties to try to share water between nations. These include the Rhine and Danube rivers in Europe, the Nile River in North Africa, the Ganges and Brahmputra rivers in Asia, and the Parana River in South America.

FIGURE 1 Kurnell desalination plant in southern Sydney provides water to the city's population. It is powered by wind energy produced in Canberra.

3.9.3 CASE STUDY: Managing water in the Nile Basin, North Africa

The Nile River is the longest river in the world at 6695 kilometres long. It flows northward through the tropics and the highlands of eastern Africa and drains into the Mediterranean Sea in north Africa. It is also home to the Sudd wetland, Lake Victoria and 17 other wetlands sites with diverse species of flora and fauna that are listed by the Ramsar Convention.

The Nile Basin (the Nile River and all its tributaries) covers an area of about 3.1 million kilometres squared (almost the same area as the Northern Territory). The Basin covers 10 per cent of the African continent.

Ten countries share the water and land in the Nile Basin: Burundi, Democratic Republic of Congo, Egypt, Eritrea, Ethiopia, Kenya, Rwanda, Sudan, Uganda and United Republic of Tanzania. More than 330 million people live in these ten countries — 160 million live within the watershed boundaries of the Nile Basin and share its water.

Water in the Nile Basin countries is used for hydro-electricity generation, town/city and industrial water supply, agriculture, fishing, recreation, transport, tourism and waste disposal. Most people earn a living in the Nile Basin through agriculture; it sustains millions of people. The topography of the Nile enables power generation, especially in Ethiopia. Hydropower is a major water user in the Nile, relying on water passing through turbines to generate electricity.

Nile Basin Sustainability Framework

There have been many initiatives and agreements since the 1990s regarding sharing the Nile's water. All have been based on a framework for cooperation and trying to ensure there is equitable use of this important resource.

Approved in 2011, the Nile Basin Sustainability Framework (NBSF) outlines the guiding principles for water resource management and development across the Nile Basin countries. This framework guides national policy and seeks to build consensus. It supports transboundary investment projects and promotes shared benefits and environmental concerns that ensure projects have long-term benefits. A significant effort was made to strengthen database and shared geographical information system (GIS) skills so that up-to-date water flow and rainfall data are available. Without the NBSF, there would be no guidance for the sustainable development or cooperation in sustainable water management and development in the Nile Basin.

FIGURE 2 The sub-basins and countries sharing the water and land of Nile River Basin

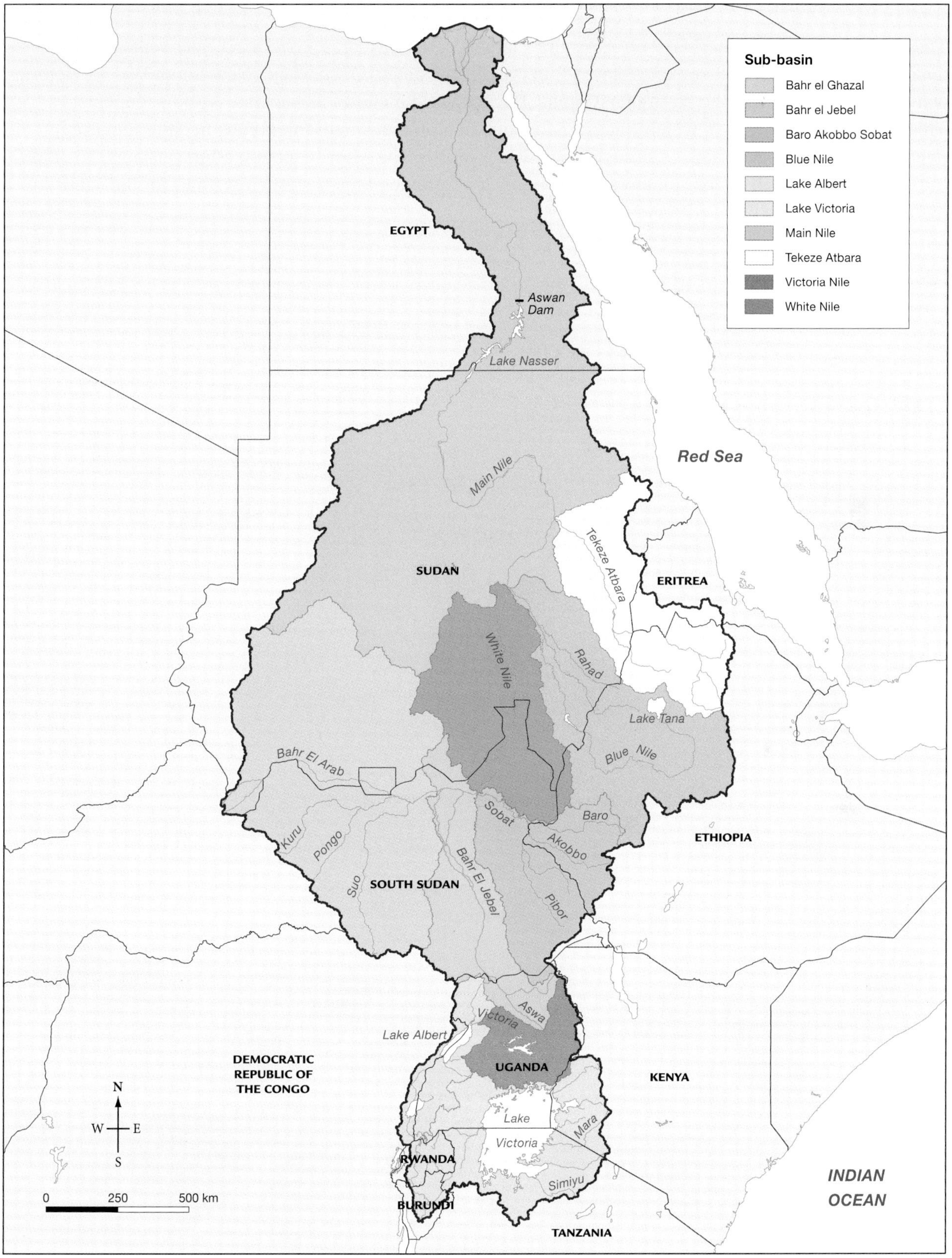

Sub-basin

- Bahr el Ghazal
- Bahr el Jebel
- Baro Akobbo Sobat
- Blue Nile
- Lake Albert
- Lake Victoria
- Main Nile
- Tekeze Atbara
- Victoria Nile
- White Nile

EGYPT

Aswan Dam

Lake Nasser

Red Sea

SUDAN

Main Nile

White Nile

Rahad

Tekeze Atbara

ERITREA

Lake Tana

Bahr El Arab

Blue Nile

Sobat

Baro

Kuru

Pongo

Akobbo

ETHIOPIA

Suo

SOUTH SUDAN

Bahr El Jebel

Pibor

Aswa

Victoria

Lake Albert

DEMOCRATIC
REPUBLIC OF
THE CONGO

UGANDA

KENYA

N
W — E
S

Lake

Victoria

Mara

RWANDA

Simiyu

0   250   500 km

BURUNDI

TANZANIA

INDIAN
OCEAN

Source: Nile Basin Initiative

3.9.4 CASE STUDY: Managing water within a country: the Murray–Darling Basin

The Murray–Darling Basin (MDB) is Australia's largest catchment area, covering 14 per cent of the country's total landmass. It stretches across four Australian states (Queensland, New South Wales, Victoria and South Australia) and the Australian Capital Territory, and includes 23 separate river catchments. This is a changed system, in which water is taken out all along its length for storage in dams and use in irrigation. During long periods of drought, little water reaches the Murray mouth, often causing it to close. The Queensland floods in the summers of 2011 and 2012 provided the largest river flows for the MDB in many years.

FIGURE 3 Map of the Murray–Darling Basin. The Australian government has been managing the basin through the Murray–Darling Basin Authority since 2008.

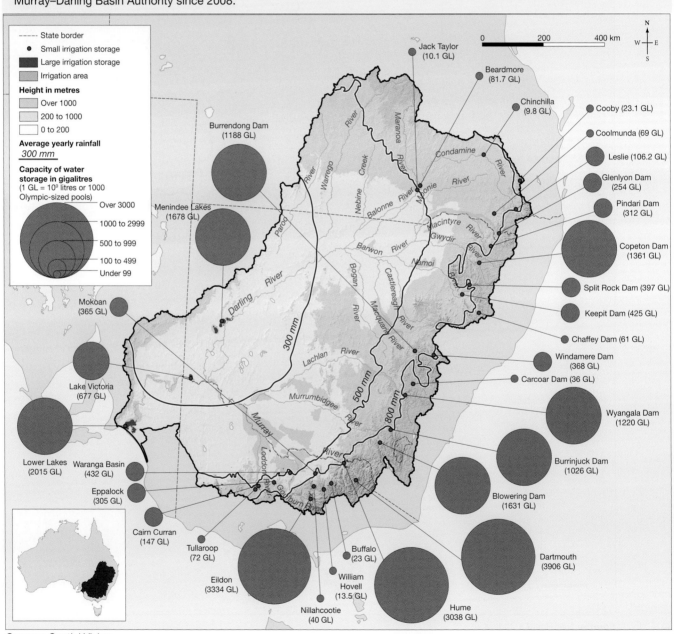

Source: Spatial Vision

The management of this river system was originally the responsibility of each of the four states and the territory, which resulted in a lot of conflict. In 2008, the Australian government took control of the MDB and prepared plans for how it should be managed.

Despite the federal government taking over management of the MDB, there has continued to be controversy over decisions made about sharing the water. In 2018–19, very low rainfall and inequitable water use in the MDB has resulted in high fish kills in the Menindee Lakes in NSW. Other impacts include reduced water allocations to farmers, who rely on water from the system for crops and stock.

FIGURE 4 Rainfall MDB measured from July 2017 to May 2018

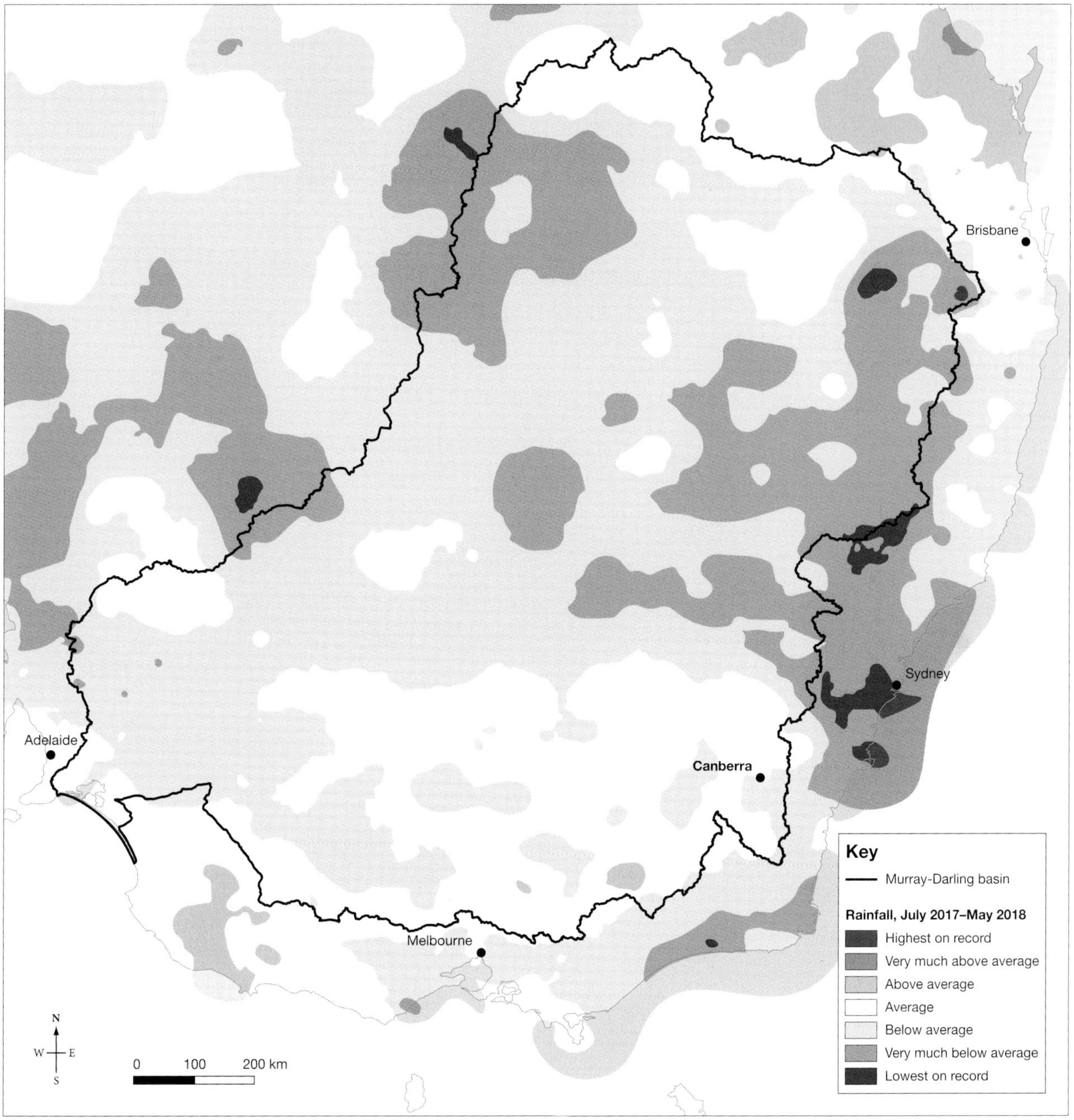
Brisbane
Sydney
Canberra
Adelaide
Melbourne
Key
Murray-Darling basin
Rainfall, July 2017–May 2018
Highest on record
Very much above average
Above average
Average
Below average
Very much below average
Lowest on record
N
W
E
S
0
100
200 km

Source: Bureau of Meteorology

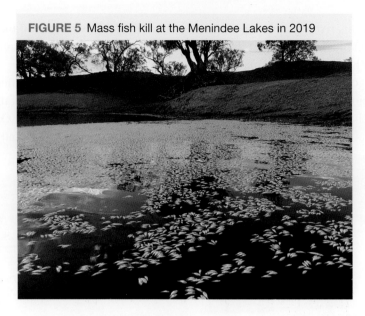

FIGURE 5 Mass fish kill at the Menindee Lakes in 2019

3.9.5 Other water management solutions for Australia

Over recent years, and especially after prolonged droughts, many different solutions have been suggested to solve our water problems. Some of these seem impractical, such as towing icebergs from Antarctica; others

FIGURE 6 Possible water management solutions

Install a watertank in every house.

Divert rivers back inland to stop water reaching the sea.

Australians should reuse and recycle water.

Fix all the leaking water pipes.

Build more desalination plants.

Use more groundwater.

Build more dams.

Possible water management solutions

Relocate most of Australia's farms to the north where most of the rain is.

Use pipes to move water from areas of high rainfall to areas of low rainfall.

Use cloudseeding to make it rain more.

Increase the price of water to homes.

Use drip irrigation on all crops to save water.

Move water from wet areas such as Tasmania and northern Australia to dry areas using ship tankers.

Cover reservoirs to stop evaporation.

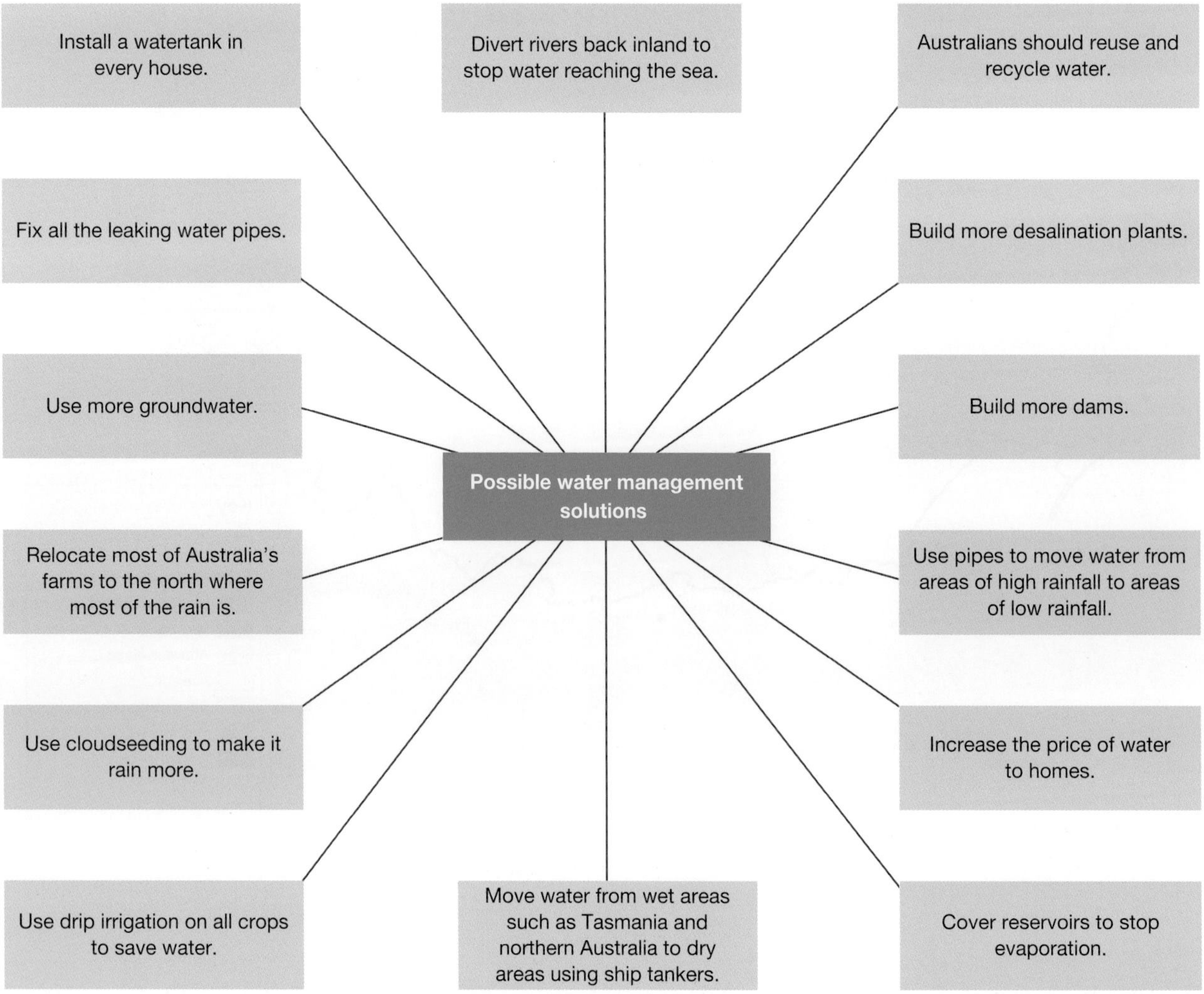

have generated much discussion, such as fixing all the leaking pipes in towns, cities and outback bores. Each of the suggestions has to be considered in light of various factors: cost, impact on people, impact on the environment, technology and politics.

3.9.6 Managing water use at home

About 50 per cent of household water in Australia is used in the garden. Inside the house, approximately 80 per cent of water used is in the shower, toilet and laundry. It is predicted that Australia's growing population will, by 2051, need nearly twice as much water as we do now.

Policy changes such as water restrictions can reduce water usage. Restrictions can include bans on hosing down driveways, washing cars with hoses and watering private lawns, and cutting back sprinkler use during the day.

Tips to save water

In the house

- Take shorter showers.
- Turn off taps firmly and fix any leaks.
- Use water-efficient shower heads.
- Install a dual-flush toilet or adaptor.
- Use water-efficient appliances and use them only when they are full.
- Keep a jug of cold water in the fridge so you don't need to run the tap until the water is cold enough to drink.
- When replacing appliances, choose water-efficient models (AAA or AAAA rating).

In the garden

- Plant local native plants, which need only rainfall.
- Group plants with similar watering needs together and water them together.
- Use mulch on the garden beds to stop soil drying out — evaporation can be reduced by up to 70 per cent this way.
- Use a trigger nozzle on your hose.
- Install a timer on your outdoor taps.
- Use trickle irrigation systems rather than sprinklers for garden beds. They direct water where it is needed, and less water is lost to evaporation and wind-drift.
- Use a pool cover when the pool is not being used.

The long drought experienced by Australia in the first decade of the twenty-first century forced many governments to offer rebates on purchases of water-saving products. Items and services included buying and installing water tanks and grey water systems; dual flush toilets; water-saving shower heads and water-efficient washing machines. People were encouraged to spend money on these in order to save water.

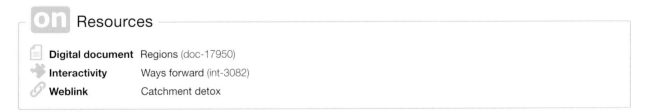

on **Resources**

📄	**Digital document**	Regions (doc-17950)
	Interactivity	Ways forward (int-3082)
🔗	**Weblink**	Catchment detox

Explore more with myWorldAtlas

Deepen your understanding of this topic with related case studies and questions.
- Investigating Australian Curriculum topics > Year 7: Water in the world > **Salisbury Council — Aquifer storage, transfer and recovery**
- Investigate additional topics > Managing water resources > **Murray–Darling Basin**

3.9 INQUIRY ACTIVITIES

1. Conduct some further research to find out how water is being managed along the Jordan River in west Asia (the Middle East). (Use the **Regions** resource in the Resources tab.) **Examining, analysing, interpreting**

2. Investigate the use of desalination in countries in west Asia. **Examining, analysing, interpreting**

3. If water were oil, leaking pipes would be fixed immediately. However, there is still a perception that water is not as valuable as oil, so the same investment is not made in it. Imagine you work for an advertising company and have to convince your audience that water is more valuable than oil. Film your advertisement and use a video editing tool to create music and voiceovers. Present this to your class and your school. **Evaluating, predicting, proposing**

4. Study the list of possible water solutions shown in **FIGURE 6** and choose the ones that are most likely to work.

   You will work in pairs to conduct some research about one of the proposed methods. There will be advantages and disadvantages for each. You will need to use the internet and libraries to find your information, and you will need to find out:

   - how the solution will be carried out, including the technology that might be used
   - which *places* in Australia are most likely to be involved and why (this might include rainfall data, Google images or maps, and photographs)
   - how much it will cost and who will pay
   - what impact the solution will have on the *environment*
   - what impact there will be on people
   - whether there are any political implications
   - where the solution will work best (shown on a map).

Presenting your information

Any data collected needs to be presented in an appropriate way. Climate data could be represented as a climate map; a satellite image or photograph could be annotated with notes; models or plans of the solution could be drawn.

Sharing your information

The class will need to share all the information found so that a decision can be made about the most viable solutions. Share via a presentation (such as a Prezi), a class wiki or blog.

Making a decision

Conduct a class vote to remove 5 of the 14 solutions immediately after the information sharing. Do this by writing each solution on a board and voting on each one. You will be left with nine water management solutions.

Diamond ranking activity

Use the remaining nine solutions to complete a diamond ranking. Make a copy of the nine solutions and write these on separate cards or sticky labels. Individually, use the information you shared about the solutions to rank the solutions from the most viable to the least viable from the top of the chart. It is often easier to work on the two extremes, top and bottom, and then continue working from there.

   Once you have your ranking, work in groups of four. Explain why you chose the ranking and see whether each of you can agree on the same ranking. Work together as a class and discuss the solutions again and see if you can arrive at a class ranking.

Conclusion

Write a short report, and include in it the final ranking of water management solutions for Australia. Include the discussions and explanations in your report.

   Now write a letter to your local state or territory water authority, outlining what you consider to be the three best management solutions for Australia and why.

Evaluating, predicting, proposing

3.9 EXERCISES

Geographical skills key: GS1 Remembering and understanding **GS2** Describing and explaining **GS3** Comparing and contrasting **GS4** Classifying, organising, constructing **GS5** Examining, analysing, interpreting **GS6** Evaluating, predicting, proposing

3.9 Exercise 1: Check your understanding

1. **GS2** Why is it difficult to manage water when the water supply crosses country or state borders?
2. **GS2** Read the case study on the Nile Basin and study **FIGURE 2**. Outline the potential difficulty in managing the water in the Nile Basin.
3. **GS3** Study **FIGURES 4, 5** and **6** and read about the management of the Murray–Darling Basin.
   (a) How does the Murray–Darling Basin compare to the Nile River Basin?
   (b) How has the recent drought affected the MDB?
4. **GS1** What is one gigalitre of water equivalent to?
5. **GS2** List all the ways that you and your family save water.

3.9 Exercise 1: Check your understanding

1. **GS5** Discuss the following statement: 'There is enough water for all purposes if it is managed well'.
2. **GS5** Study **FIGURE 5** and answer the following questions:
   (a) Name the three largest storage dams in the Murray–Darling Basin (MDB).
   (b) Where is each located?
   (c) How many 1000–2999 GL storages are located in NSW?
   (d) Why does so little water reach the Murray River mouth in South Australia?

Try these questions in learnON for instant, corrective feedback. Go to www.jacplus.com.au.

3.10 Thinking Big research project: Desalination plant advertising

SCENARIO

It is 2025 and the drought that began in 2019 has continued without foreseeable sufficient rain. Your media company has been asked to develop a suite of advertisements to explain to Victorians the need to spend government money on a second desalination plant.

Select your learnON format to access:
- the full project scenario
- details of the project task
- resources to guide your project work
- an assessment rubric.

 Resources

projectsPLUS Thinking Big research project: Desalination plant advertising (pro-0234)

3.11 Review

3.11.1 Key knowledge summary

Use this dot point summary to review the content covered in this topic.

3.11.2 Reflection

Reflect on your learning using the activities and resources provided.

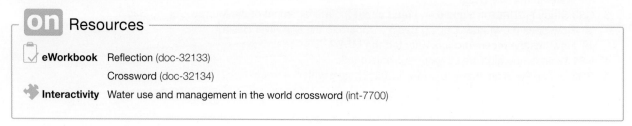

on Resources

☑ **eWorkbook** Reflection (doc-32133)

Crossword (doc-32134)

🧩 **Interactivity** Water use and management in the world crossword (int-7700)

KEY TERMS

aquifer a body of permeable rock below the Earth's surface which contains water, known as groundwater. Water can move along an aquifer.

biodiversity the variety of life in the world or in a particular habitat or ecosystem

blue water the water in freshwater lakes, rivers, wetlands and aquifers

climate change a change in the world's climate. This can be very long term or short term, and can be caused by human activity.

discharge the volume of water that flows through a river in a given time

drainage basin the entire area of land that contributes water to a river and its tributaries

evaporation the process by which water is converted from a liquid to a gas and thereby moves from land and surface water into the atmosphere

green water water that is stored in the soil or that stays on top of the soil or in vegetation

improved drinking water drinking water that is safe for human consumption

precipitation rain, sleet, hail, snow and other forms of water that falls from the sky when water particles in clouds become too heavy

rainfall variability the change from year to year in the amount of rainfall in a given location

relative humidity the amount of moisture in the air

run-off precipitation not absorbed by soil, and which runs over the land and into streams

virtual water all the water used to produce goods and services. Food production uses more water than any other production.

water footprint the total volume of fresh water that is used to produce the goods and services consumed by an individual or country

water scarcity a situation that occurs when the demand for water is greater than the supply available

water stress a situation that occurs in a country with less than 1000 cubic metres of renewable fresh water per person

4 Water variability and its impacts

4.1 Overview

From droughts to floods to heat waves to blizzards. Why is weather so unpredictable?

4.1.1 Introduction

Every person on the planet interacts with the weather on a daily basis. Sometimes we feel hot, sometimes we feel cold. The constant change in the weather also affects the environment.

Weather influences the level of precipitation experienced in different places. If there is too little rain, drought can develop, sometimes producing heatwaves — days of dry, hot weather. If there is too much rain, flooding will occur. These extreme weather events have many effects, which can be more severe in certain parts of the world. While we cannot control these events, we can learn from the past and plan to minimise their impact in the future.

LEARNING SEQUENCE

4.1 Overview
4.2 Weather and climate
4.3 **SkillBuilder:** Reading a weather map
4.4 Natural hazards and natural disasters
4.5 Drought in Australia
4.6 Managing dry periods to reduce the impacts of drought
4.7 Bushfires in Australia
4.8 Floods and floodplains
4.9 Managing the impacts of floods
4.10 **SkillBuilder:** Interpreting diagrams
4.11 **Thinking Big research project:** Water access comparison
4.12 **Review**

To access a pre-test and starter questions, and receive immediate, **corrective feedback** and **sample responses** to every question, select your learnON format at www.jacplus.com.au.

4.2 Weather and climate

4.2.1 How does weather change?

Our Earth is surrounded by a band of gases called the atmosphere. It protects our planet from the extremes of the sun's heat and the chill of space, making conditions just right for supporting life. The atmosphere has five different layers. The layer that starts at ground level and ends about 16 kilometres above Earth is called the troposphere. Our weather is the result of constant changes to the air in the troposphere. These changes sometimes cause extreme weather events.

Droughts, **floods**, cyclones, tornadoes, heatwaves and snowfalls — even cloudless days with gentle breezes — all begin with changes to the air in the troposphere. The five main layers in the Earth's atmosphere all differ from one another. For example, the troposphere contains most of the **water vapour** in the atmosphere. As a result, this layer has an important link to **precipitation**.

All weather conditions result from different combinations of three factors: air temperature, air movement and the amount of water in the air. The sun influences all three.

FIGURE 1 Australia experiences a diversity of weather, which has a major effect on how we live.

First, the sun heats the air. It also heats the Earth's surface, which in turn heats the air even more. How hot the Earth's surface becomes depends on the season and the amount of cloud cover.

Second, the sun causes air to move. This is because the sun heats land surfaces more than it heats the oceans. As the warm air over land gets even warmer, it expands and rises. When hot air rises, colder air moves in to take its place.

Third, the sun creates moisture in the air. The heat of the sun causes water on the Earth's surface to **evaporate**, forming water vapour. As this water vapour cools, it condenses, forming clouds. It may return to Earth as rain, dew, fog, snow or hail.

At times these three factors — temperature, air movement and water vapour — can create extreme weather events. Very high air temperatures influence heatwaves; rapidly rising air plays a part in the formation of cyclones; and excess rain can create flooding.

4.2.2 The difference between weather and climate

Weather is the day-to-day, short-term change in the atmosphere at a particular location. Extreme weather events are often described as unexpected, rare or not fitting the usual pattern experienced at a location.

Climate is the average of weather conditions that are measured over a long time. Places that share the same type of weather are said to lie in the same climatic zone. Because of the size of the Australian continent, its climate varies considerably from one region to another.

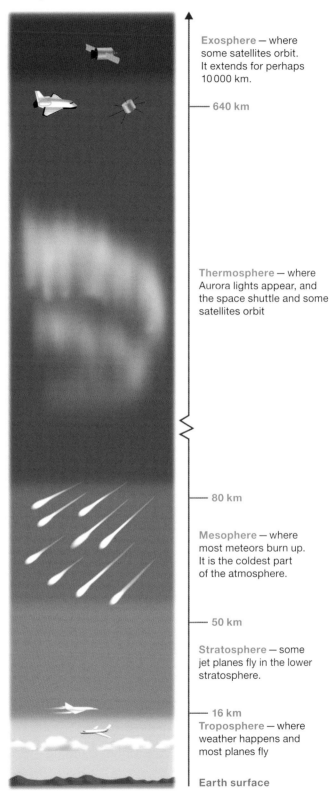

FIGURE 2 Structure of the Earth's atmosphere (not to scale)

Exosphere — where some satellites orbit. It extends for perhaps 10 000 km.

640 km

Thermosphere — where Aurora lights appear, and the space shuttle and some satellites orbit

80 km

Mesosphere — where most meteors burn up. It is the coldest part of the atmosphere.

50 km

Stratosphere — some jet planes fly in the lower stratosphere.

16 km

Troposphere — where weather happens and most planes fly

Earth surface

FIGURE 3 Climatic zones of Australia

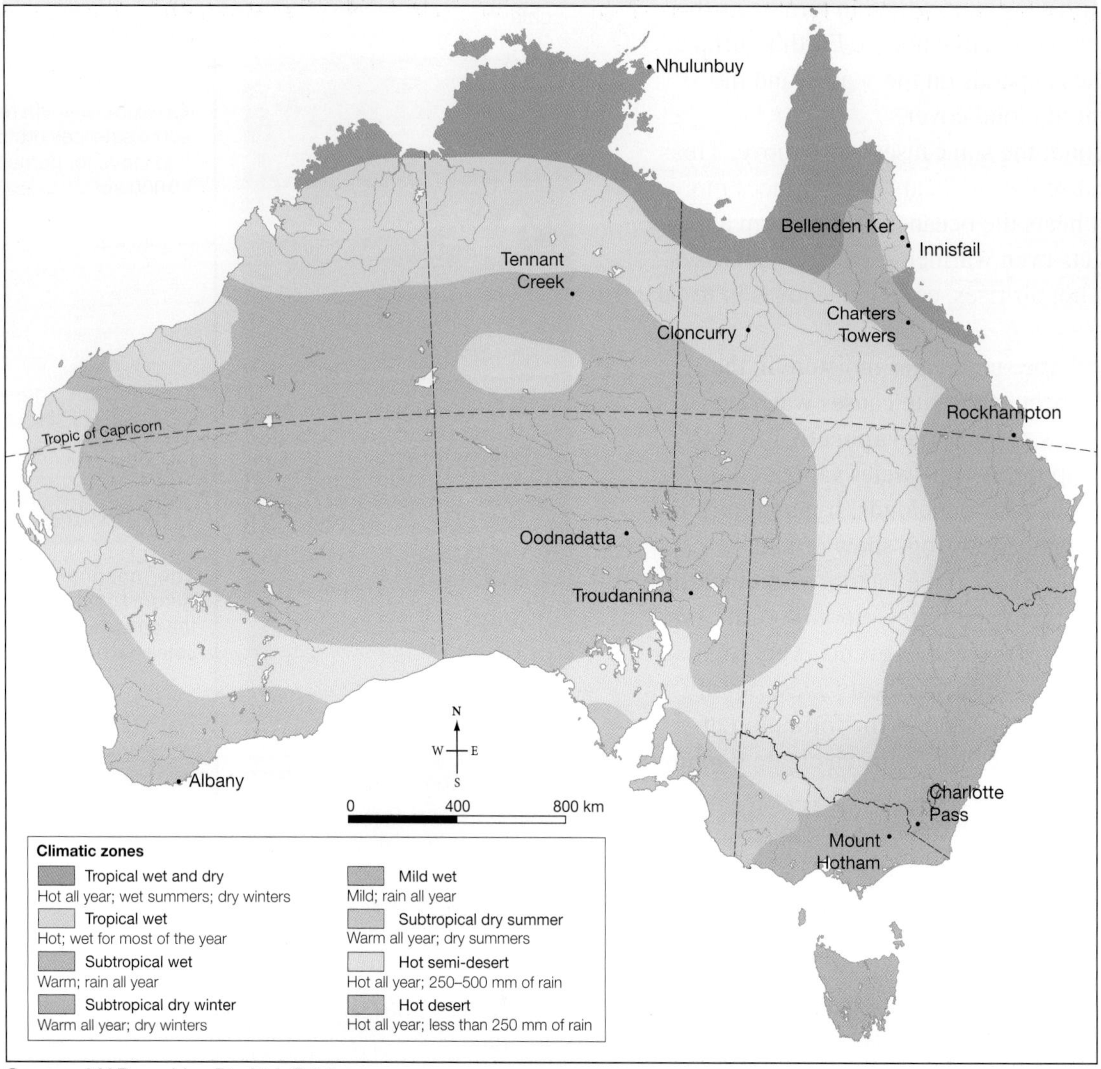
Nhulunbuy
Bellenden Ker
Innisfail
Tennant Creek
Cloncurry
Charters Towers
Rockhampton
Tropic of Capricorn
Oodnadatta
Troudaninna
N
W
E
S
0
400
800 km
Albany
Charlotte Pass
Mount Hotham
Climatic zones
Tropical wet and dry
Hot all year; wet summers; dry winters
Tropical wet
Hot; wet for most of the year
Subtropical wet
Warm; rain all year
Subtropical dry winter
Warm all year; dry winters
Mild wet
Mild; rain all year
Subtropical dry summer
Warm all year; dry summers
Hot semi-desert
Hot all year; 250–500 mm of rain
Hot desert
Hot all year; less than 250 mm of rain

Source: MAPgraphics Pty Ltd, Brisbane

4.2 INQUIRY ACTIVITY

In a magazine or newspaper or online, find a photograph that shows an example of one type of weather. Paste the picture in the centre of a page and add labels about the impact of that weather on the ***environment*** (for example, creating puddles) and on what we do (such as the clothes people wear).

Examining, analysing, interpreting

4.2 EXERCISES

Geographical skills key: GS1 Remembering and understanding **GS2** Describing and explaining **GS3** Comparing and contrasting **GS4** Classifying, organising, constructing **GS5** Examining, analysing, interpreting **GS6** Evaluating, predicting, proposing

4.2 Exercise 1: Check your understanding

1. **GS1** What is the name of the layer of the atmosphere where all Earth's weather happens?
2. **GS1** Define the term 'troposphere'.
3. **GS1** In which levels of the atmosphere are the following features found?
   (a) Most passenger planes
   (b) Orbiting satellites
   (c) Burning meteors
   (d) The Aurora lights

4. **GS2** Explain the difference between weather and climate.
5. **GS2** Draw three diagrams to help you explain the factors that influence the following weather conditions:
   (a) air temperature
   (b) air movement
   (c) the amount of water in the air.

4.2 Exercise 2: Apply your understanding

1. **GS3** Look carefully at the photographs in **FIGURE 1**.
   (a) Describe the weather event in each photograph.
   (b) How would each weather event affect people's lives?
2. **GS6** Look carefully at the map of Australia's climatic zones in **FIGURE 3**. Predict which two settlements, or **places**, might be at risk of flood. Make sure you explain why you chose them.
3. **GS6** Look at the **environment** outside the window.
   (a) What is the weather like? Do you think it matches the climatic zone in which you live? Explain.
   (b) Now check to see your climatic zone using **FIGURE 3**. If your answers are different, explain why this may have occurred.
4. **GS2** Describe how the weather affected you yesterday.
5. **GS3** Describe the relationship between Australia's climate and the places where we choose to live.

Try these questions in learnON for instant, corrective feedback. Go to www.jacplus.com.au.

4.3 SkillBuilder: Reading a weather map

What are weather maps?

Weather maps, or synoptic charts, show weather conditions over a larger area at any given time. They appear every day in newspapers and on television news. Being able to read a weather map is a useful skill because weather affects our everyday life.

Select your learnON format to access:

- an overview of the skill and its application in Geography (Tell me)
- a video and a step-by-step process to explain the skill (Show me)
- an activity and interactivity for you to practise the skill (Let me do it)
- questions to consolidate your understanding of the skill.

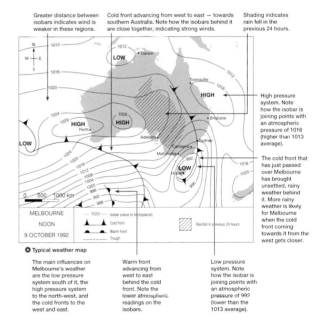

Greater distance between isobars indicates wind is weaker in these regions.

Cold front advancing from west to east — towards southern Australia. Note how the isobars behind it are close together, indicating strong winds.

Shading indicates rain fell in the previous 24 hours.

High pressure system. Note how the isobar is joining points with an atmospheric pressure of 1016 (higher than 1013 average).

The cold front that has just passed over Melbourne has brought unsettled, rainy weather behind it. More rainy weather is likely for Melbourne when the cold front coming towards it from the west gets closer.

MELBOURNE
NOON
9 OCTOBER 1992

1020 — Isobar (value in hectopascal)
Cold front
Warm front
Trough
Rainfall in previous 24 hours

Typical weather map

The main influences on Melbourne's weather are the low pressure system south of it, the high pressure system to the north-west, and the cold fronts to the west and east.

Warm front advancing from west to east behind the cold front. Note the lower atmospheric readings on the isobars.

Low pressure system. Note how the isobar is joining points with an atmospheric pressure of 992 (lower than the 1013 average).

Resources

Video eLesson Reading a weather map (eles-1637)

Interactivity How to read a weather map (int-3133)

4.4 Natural hazards and natural disasters

4.4.1 What are natural hazards?

Australia is prone to a wide variety of **natural hazards**, which range from drought and bushfire to flooding. Many of these events are part of the weather's natural cycle. However, human actions such as overgrazing, deforestation and the alteration of natural waterways have sometimes increased the impact of these hazards. So why do people continue to live in areas that are at risk of experiencing these hazards?

There is a difference between natural hazards and **natural disasters**. A hazard is an event that is a *potential* source of harm to a community. A disaster occurs as the result of a hazardous event that dramatically affects a community. There are four broad types of natural hazard:

1. atmospheric — for example, cyclones, hailstorms, blizzards and bushfires
2. hydrological — for example, flooding, wave action and glaciers
3. geological — for example, earthquakes and volcanoes
4. biological — for example, disease epidemics and plagues.

Natural hazards that are linked to the weather are categorised into the atmospheric and hydrological types. Hazards such as flooding and cyclones could also be termed extreme weather events.

Some natural hazards are influenced by the actions of people and where they choose to locate themselves. For example, the severity of a flood depends not only on the amount and duration of rainfall that occurs. Humans can influence floods by building on floodplains and not planning well for disaster. Environmental degradation and poor urban planning can also turn natural hazards into natural disasters.

FIGURE 1 The link between vulnerability and disaster

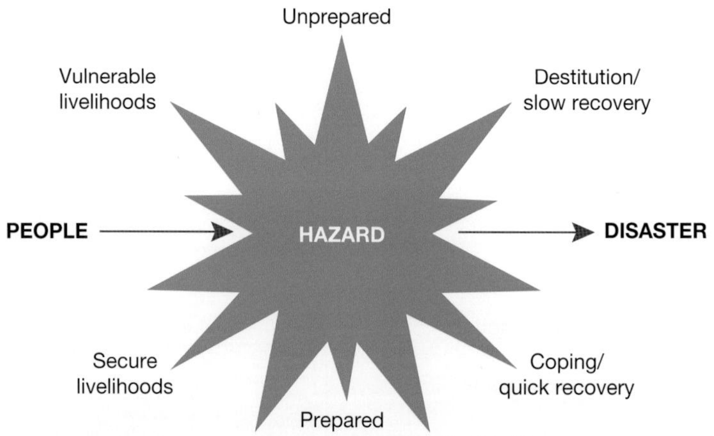

Five of Australia's costliest natural disasters

- *Drought*, Australia-wide yet mainly in New South Wales, 2018: $12 billion cost
- *Flood*, Queensland, New South Wales and Victoria 2010–2011: 35 deaths, 20 000 homes destroyed in Brisbane alone, $5.6 billion cost
- *Cyclone*, Cyclone Tracy, Darwin 1974: 65 deaths, 10 800 buildings destroyed, $4.18 billion cost
- *Earthquake*, Newcastle 1989: 13 deaths, 50 000 buildings damaged, more than $4 billion cost
- *Bushfire*, Black Saturday, Victoria 2009: 173 deaths, 3500 buildings destroyed, $1.5 billion cost

Top five casualty rate natural disasters worldwide since 2000

- *Earthquake*, Haiti, 2010: estimated range 85 000–316 000 deaths
- *Tsunami*, Indian Ocean 2004: approximately 230 000 deaths
- *Cyclone*, Cyclone Nargis, Myanmar, 2008: at least 146 000 deaths
- *Earthquake*, Sichuan, China, 2008: approximately 87 400 deaths
- *Earthquake*, Kashmir, Pakistan, 2005: approximately 79 000 deaths

FIGURE 2 Australia's natural hazards and disasters

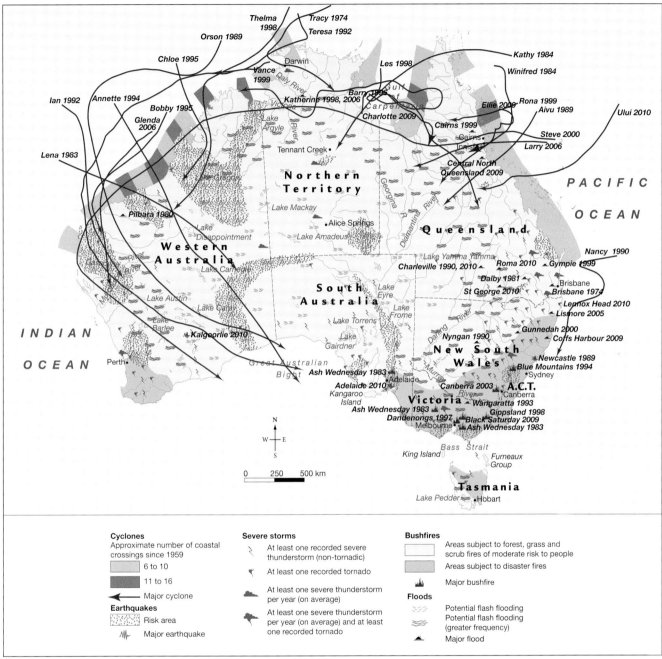

Source: MAPgraphics Pty Ltd, Brisbane

Risk is the possibility of negative effects caused by a natural hazard. Therefore, the type of hazard experienced, along with the **vulnerability** of the people affected, will determine the risk faced. The poorest people in the world are vulnerable because their ability to recover from the impact of a hazard is hampered by their lack of resources. In an event such as a flood or earthquake, people lose their personal belongings, homes and livestock, which are often linked to their incomes, continuing the cycle of poverty. However, in regions that are adequately prepared, and where there is support to cope and rebuild, people recover more quickly.

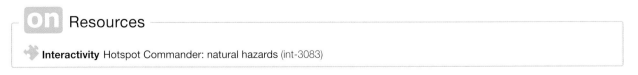

Resources

Interactivity Hotspot Commander: natural hazards (int-3083)

4.4 EXERCISES

Geographical skills key: GS1 Remembering and understanding **GS2** Describing and explaining **GS3** Comparing and contrasting **GS4** Classifying, organising, constructing **GS5** Examining, analysing, interpreting **GS6** Evaluating, predicting, proposing

4.4 Exercise 1: Check your understanding

1. **GS3** How do natural hazards and natural disasters differ?
2. **GS2** Explain how a flood is both a natural and human hazard.
3. **GS2** Explain why the risk of experiencing a natural disaster depends on the geographical location of a community.
4. **GS2** Describe key *changes* that natural hazards and natural disasters can cause to an *environment*.
5. **GS1** Match the following natural disasters to the correct details below.

Earthquake	Darwin, 1974: 65 deaths, 10 800 buildings destroyed, $4.18 billion cost
Drought	Haiti, 2010: estimated range 85 000–316 000 deaths
Cyclone Nargis	Australia-wide yet mainly in New South Wales, 2018: $12 billion cost
Flood	Myanmar, 2008: at least 146 000 deaths
Cyclone Tracy	Queensland, New South Wales and Victoria 2010–2011: 35 deaths, $5.6 billion cost

4.4 Exercise 2: Apply your understanding

1. **GS5** Refer to **FIGURE 2**.
   (a) What types of natural disasters occur most often in Australia?
   (b) Describe the location of Australia's cyclone hazard zone.
   (c) Give one example of a *place* that has suffered a bushfire disaster.
   (d) What type of hazards are *places* around Newcastle subject to?
   (e) What would be the likely impact of a large earthquake occurring in the earthquake risk area in the Northern Territory?
2. **GS6** The casualty rate for the Haitian earthquake is heavily debated. Suggest possible reasons for a lack of accuracy in this case.
3. **GS6** Is your local area at risk of any natural disasters? If so, identify the category of disaster which is most likely to affect your area.
4. **GS3** Explain the difference between risk and impacts in reference to natural disasters.
5. **GS6** What role can Australia play in supporting countries in our region that experience frequent natural disasters?

Try these questions in learnON for instant, corrective feedback. Go to www.jacplus.com.au.

4.5 Drought in Australia

4.5.1 What is a drought?

Australia is the driest inhabited continent on Earth. The main reason Australia is so dry is that much of the continent lies in an area dominated by high atmospheric pressure for most of the year, which brings dry, stable, sinking air to the country. Australia also experiences great variation in its rainfall due to the **southern oscillation** and **El Niño**.

Low average rainfall and extended dry spells are a normal part of life throughout most of Australia. The continent is located in a zone of high pressure that creates conditions of clear skies and low rainfall. Drought conditions occur when the high-pressure systems are more extensive than usual, creating long or severe rainfall shortages. A drought is a prolonged period of below-average rainfall, when there is not enough water to supply our normal needs. Because people use water in so many different ways and in such different quantities, there is no universal amount of rainfall that defines a drought.

The term *drought* should not be confused with low rainfall. Sydney could experience a drought and have more rainfall during that period than Alice Springs, which could be experiencing above average rainfall. If low rainfall meant drought, then much of Australia would be in drought most of the time.

Droughts affect all parts of Australia over a period of time. Some droughts can be localised while other parts of the country receive good rain. Droughts can be short and intense, such as the drought that lasted from April 1982 to February 1983; or they can be long-lived, such as the 2002–2009 drought.

Different weather systems affect different parts of Australia, so there is little chance that all of Australia would experience drought at the same time.

FIGURE 1 Weather events in (a) a typical year and (b) an El Niño year

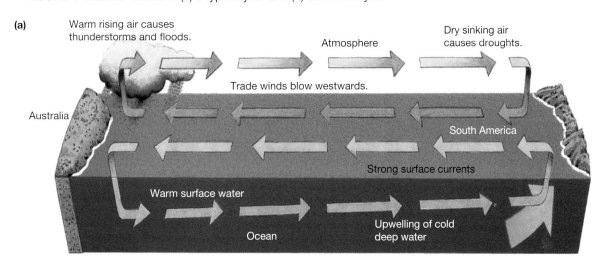

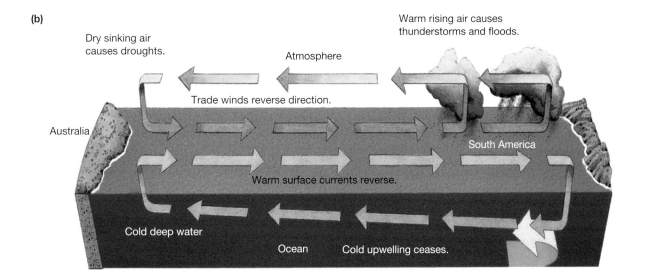

4.5.2 El Niño

Australian droughts are heavily influenced by what **meteorologists** call an El Niño event. In a normal year, warm surface water is blown west across the Pacific Ocean towards Australia. This brings heavy rain to northern Australia, Papua New Guinea and Indonesia. On the other side of the Pacific, South America experiences drought. When there is an El Niño event, these winds and surface ocean currents reverse their direction. The warm, moist air is pushed towards South America. This produces rain in South America and drought in Australia.

4.5.3 Southern oscillation

Fluctuations in rainfall have several causes that are not fully understood. Probably the main cause of major rainfall fluctuations in Australia is the southern oscillation, which is a major air pressure shift between the Asian and east Pacific regions. The strength and direction of the southern oscillation is measured by a simple index called the southern oscillation index (SOI). The SOI is calculated from monthly or seasonal fluctuations in air pressure between Tahiti and Darwin.

In an average rainfall year with 'typical' pressure patterns, the SOI is between −10 and +10. If the SOI is strongly negative (below −10), this means that the air pressure at sea level in Darwin is higher than in Tahiti, and an El Niño event occurs.

During an El Niño event, there is less than average rainfall over much of Australia. During this period, drought will occur. If the SOI becomes strongly positive (above +10), this means that the air pressure in Darwin is much lower than normal and a La Niña event occurs. During this period, above average rainfall will occur.

In recent years, scientists have made great advances in understanding and forecasting El Niño and southern oscillation events. The National Climate Centre in Australia produces outlooks on rainfall three months ahead. These outlooks are proving to be of great value to farmers and especially valuable for ecologically sustainable development in rural areas.

FIGURE 2 Areas affected by El Niño

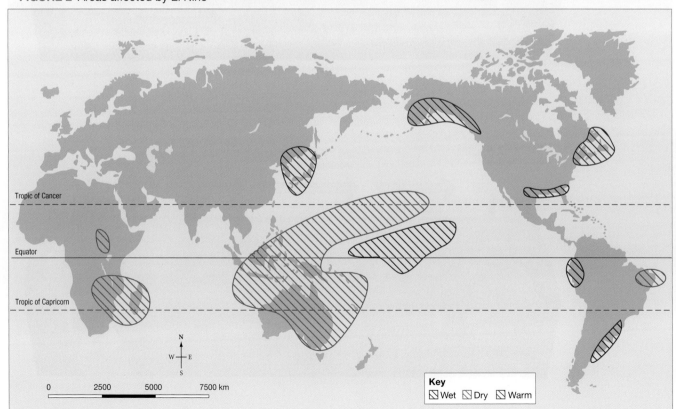

Source: MAPgraphics Pty Ltd, Brisbane

4.5.4 Impacts of drought: Australia, 2002–2009

Around the world, droughts often develop due to similar weather conditions. However, their impacts on different communities that live around the world can be diverse. Often, developed countries find themselves economically worse off during a drought period. In contrast, developing countries usually face devastating social consequences as a result of an extremely dry weather period.

Droughts can last for many years. They may be widespread or confined to small areas. The drought that started in 2002 affected large areas of Australia by 2005 (see **FIGURE 3**). When Australia experiences a drought, agriculture suffers first and most severely, but eventually everyone feels the impact.

Due to the severe lack of water caused by the drought, many farmers faced production losses because they were not able to sustain their crops or sufficiently feed their livestock. This had negative **economic** impacts.

- By 2004, dairy farmers had experienced a 4.5 per cent drop in their incomes.
- Cotton crops were devastated by the shortage of water.
- Up to 20 cotton communities and approximately 10 000 people in the industry were affected.
- Some communities had to cut production by 60 to 100 per cent.
- Cattle and sheep farmers found it hard to find stockfeed, and prices increased. As a result, herds grew smaller.

FIGURE 3 Extent of the drought — changes in vegetation cover, 2005

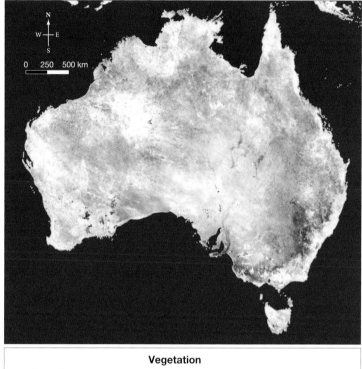

Vegetation		
Increasing	No change	Decreasing
1.0	0.0	−1.0

Source: Spatial Vision

In rural towns, jobs were lost and many businesses failed. Some people found themselves forced to leave drought-affected areas in search of other work. Many never returned. Very long droughts cause country people much heartache, and this can result in the break-up of families. It can also lead to severe depression in some individuals. However, the Australian government set up a fund that farmers and people in agricultural businesses can apply to for financial relief when their incomes are disrupted by drought. Counselling hotlines are also available to offer support.

Along with these economic and social impacts, the Australian environment suffers in drought. Droughts have a bad effect on topsoil in Australia. During drought conditions, millions of tonnes of topsoil may be blown away (see **FIGURE 4**). This loss takes many years to replace naturally, if it is ever replaced. The loss of topsoil can make many regions far less productive, making it harder for farmers to recover once the drought has broken.

FIGURE 4 'Dust' (topsoil) blown from drought–affected inland Australia blankets Sydney, 23 September 2009.

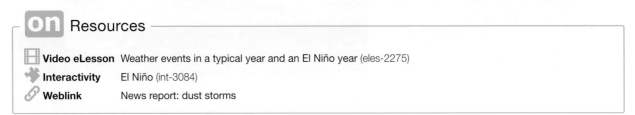

on Resources

Video eLesson	Weather events in a typical year and an El Niño year (eles-2275)	
Interactivity	El Niño (int-3084)	
Weblink	News report: dust storms	

4.5 INQUIRY ACTIVITIES

1. List some of the short-term effects that drought can have on Australia. Once you have listed them, try to come up with some long-term impacts that Australia and its people would experience if these short-term impacts continued for up to ten years. **Evaluating, predicting, proposing**

2. List all the **environmental**, economic and social impacts of drought. Using this list, create a flow diagram to illustrate how these three impact groups relate to, connect to and influence each other. Use this flow diagram to get you started. You can add more boxes and arrows to show how elements are connected. **Classifying, organising, constructing**

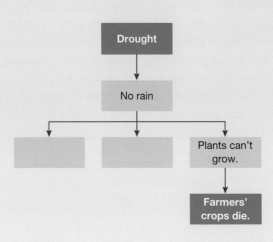

4.5 EXERCISES

Geographical skills key: GS1 Remembering and understanding **GS2** Describing and explaining **GS3** Comparing and contrasting **GS4** Classifying, organising, constructing **GS5** Examining, analysing, interpreting **GS6** Evaluating, predicting, proposing

4.5 Exercise 1: Check your understanding

1. **GS1** What is drought?
2. **GS1** Refer to **FIGURE 1**, which compares conditions during a typical year and an El Niño year, and study the text. Use the following words to complete the sentences below: stable, moist, cooler, east, drought, Tahiti, dry, warm, north, Darwin.

   During an El Niño event, the normally _____ sea in the oceans to the _____ and _____ of Australia are replaced by much _____ water. The air pressure in _____ begins to fall relative to the air pressure in _____. The normal _____ easterly trade winds change their direction. The result is _____ and _____ air and severe_____.
3. **GS2** Why is Australia so dry?
4. **GS2** What is the SOI and how is it calculated?
5. **GS2** What do the following SOIs indicate?
   (a) Between +10 and −10
   (b) >+10
   (c) <−10

4.5 Exercise 2: Apply your understanding

1. **GS2** Identify three key *changes* drought brings to the *environment*.
2. **GS6** Why is there little chance that all of Australia would be affected by drought at the same time?
3. **GS5** Refer to **FIGURE 2** showing the areas affected by El Niño. Describe the areas that become (a) wetter, (b) drier and (c) warmer during an El Niño event.
4. **GS2** Why is El Niño the result of the *interconnection* that occurs between Australia and South America?
5. **GS2** Describe the *scale* of the 2002–2009 drought.

Try these questions in learnON for instant, corrective feedback. Go to www.jacplus.com.au.

4.6 Managing dry periods to reduce the impacts of drought

4.6.1 Options for managing dry periods

During times of extreme water shortages, governments, communities and individuals often attempt to ensure there is a reliable water supply. The 2002–2009 drought in Australia sparked many different water-saving actions. However, in an environment prone to drought, with increasing demand for water, it is vital to protect and manage water resources at all times — not only during dry periods.

4.6.2 Option 1: government action

The Queensland Government developed the South-East Queensland (SEQ) water grid in order to secure alternative sources of water in an environment that seemed to be growing drier. Although the project began back in 2004, it remains an excellent example how the impacts of drought can be managed. This strategy aimed to connect the water sources of the region through a pipe network that could move water to different areas and thus meet the needs of local communities. The grid includes existing dams, three water treatment plants and a desalination plant, all connected by approximately 450 kilometres of pipes.

In 2008, the Western Corridor Recycled Project was completed at a cost of $2.5 billion. This project is part of the SEQ water grid and is the largest recycled water scheme in Australia. The project will supply up to 230 megalitres per day of recycled water to industry and power plants. The water also has the potential to be used by farmers and to top up drinking supplies. However, these last two uses of recycled water have created wide debate among communities.

FIGURE 1 The desalination plant at Tugun produces drinking water for south-east Queensland.

The desalination plant at Tugun on the Gold Coast can provide up to 133 megalitres of drinking water per day. Essentially this project produces drinking water by removing salts and other minerals from sea water. This technology is very successful and has been used in other regions of Australia for years, including Coober Pedy, where desalination is used to treat bore water. Internationally, there are approximately 7500 plants in operation. These desalination plants enable safe drinking water to be produced without having to rely on rainfall.

Water measurements
- 1 ML = 1 megalitre
- 1 megalitre = 1 000 000 litres
- 10 megalitres = 4 Olympic-size swimming pools

Stages of the desalination process
1. Sea water is piped from the ocean through a submerged inlet tunnel to the plant.
2. At the pre-treatment stage, particles in the sea water are micro-filtered, the pH is adjusted, and an inhibitor is added to control the build-up of scale in pipelines and tanks.
3. The sea water is forced through layers of membrane to remove salt and minerals. Concentrated salt water is separated and returned to the ocean.
4. During post-treatment, small amounts of lime and carbon dioxide are added to the water, along with chlorine for disinfection.
5. The desalinated water is blended with other Gold Coast water supplies and joins south-east Queensland's water grid to supply homes and industry.

Based on information from www.watersecure.com.au

In times of drought, governments may introduce water restrictions to limit the pressure placed on water supplies by individual households and businesses. They may also introduce **incentive** schemes that provide a **rebate** on water-saving devices, such as water tanks, which help relieve the strain on the water supply.

FIGURE 2 The SEQ water grid

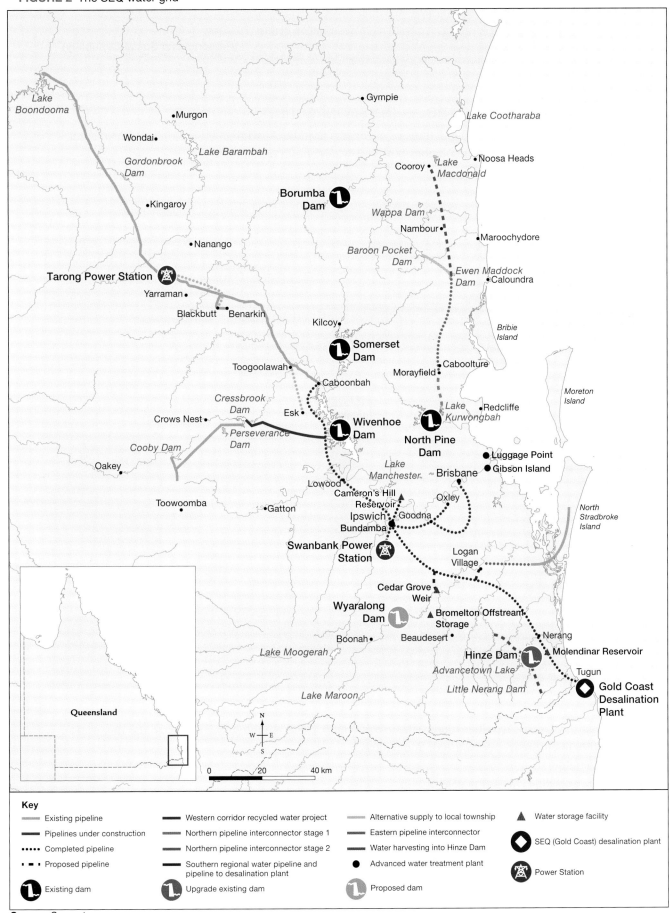

Key

Existing pipeline

Pipelines under construction

Completed pipeline

Proposed pipeline

Existing dam

Western corridor recycled water project

Northern pipeline interconnector stage 1

Northern pipeline interconnector stage 2

Southern regional water pipeline and pipeline to desalination plant

Upgrade existing dam

Alternative supply to local township

Eastern pipeline interconnector

Water harvesting into Hinze Dam

Advanced water treatment plant

Proposed dam

Water storage facility

SEQ (Gold Coast) desalination plant

Power Station

Source: Seqwater

DISCUSS

'The best way to drought-proof Australia is to build more desalination plants.' Discuss.

[Personal and Social Capability]

4.6.3 Option 2: You and me — personal action

During 2016–17, an estimated 1900 gigalitres of water was used by Australian households. With so much water used in our homes, it seems an obvious place for water conservation to begin.

FIGURE 3 Collecting rainwater from your roof can provide water for flushing toilets, watering the garden or washing the car.

Many of our day-to-day activities require the use of water. There are things we can do to use this water more efficiently to ensure it is not wasted. Some ideas include:

- putting aerators on taps
- using a hose with a shut-off nozzle
- cleaning driveways and paths with a broom rather than a hose.

FIGURE 4 shows some other ways we can use water wisely. Personal action can have a big impact. If every individual seeks to minimise wastage, significant water savings can be achieved.

FIGURE 4 ABC

Ensure your next washing machine has lots of water-efficiency stars.

Install a dual-flush toilet.

Don't run the tap when brushing Your teeth.

Ensure you completely fill your dishwasher before using it.

Dispose of tissues in the bin — don't flush them down the toilet.

Use a water-saving showerhead and keep a bucket in the shower for excess water to use on the garden.

Water the garden in the early morning or evening to reduce evaporation.

Don't keep the tap running when washing fruit and vegetables. Wash them in a bowl instead.

Have short showers. Try for three minutes!

Cover soil in mulch to retain moisture in soil. Grow drought-tolerant plants.

4.6 EXERCISES

Geographical skills key: GS1 Remembering and understanding **GS2** Describing and explaining **GS3** Comparing and contrasting **GS4** Classifying, organising, constructing **GS5** Examining, analysing, interpreting **GS6** Evaluating, predicting, proposing

4.6 Exercise 1: Check your understanding

1. **GS1** Complete the following sentence.
   Desalination is a process that removes _____ from _____.
2. **GS1** Define the following terms.
   (a) Incentive
   (b) Rebate
3. **GS1** Identify two ways the government attempts to ensure there is a reliable water supply.
4. **GS1** What is the SEQ water grid made up of?
5. **GS1** How many litres are there in 230 megalitres?

4.6 Exercise 2: Check your understanding

1. **GS2** What is the aim of a desalination plant? How does it achieve this?
2. **GS6** Will the SEQ water grid be effective in managing water during a drought period? Why or why not?
3. **GS6** Why do you think the topic of using recycled water can create debate in the community? Create a list of pros and cons for the use of recycled water.
4. **GS2** Describe two *sustainable* ways of reducing the impacts of drought. Give reasons for your choices.
5. **GS6** What geographic characteristics does an area need in order for a desalination plant to be a legitimate solution to drought?

Try these questions in learnON for instant, corrective feedback. Go to www.jacplus.com.au.

4.7 Bushfires in Australia

4.7.1 Australia's bushfire history

A bushfire is a fire that burns in grass, scrub, forest or a combination of these. They are one of the most common weather hazards faced by Australians. Along with floods and droughts, they are part of the way our natural environment functions. Unless quickly controlled, the effects of bushfires can be widespread. They can cost lives, destroy property and devastate forests and farmland.

Bushfires are not a recent occurrence in Australia; they have influenced the development of the Australian landscape for many thousands of years. Fires are an essential element in some Australian ecosystems. Many ecosystems need the intense heat created by bushfires to release seeds from plants and to replenish growth.

Indigenous Australians traditionally used fire to help them when hunting. It is believed that this use of fire contributed to the development of an open woodland ecosystem in parts of south-eastern Australia.

Early European colonisers also used fire to help clear land for crops and to remove stubble left after harvesting crops. Laws have now been passed that restrict the lighting of fires for clearing. This has led to denser vegetation in many rural areas and more leaf and bark litter on the ground, which can provide a significant amount of fuel for bushfires if they do start.

FIGURE 1 Fire facts

A Dry conditions caused by drought, searing temperatures and strong, hot northerly winds cure the bush, making it so dry that a spark can ignite a major bushfire. Grasses die off and the soil is easily blown away.

B Many animals perish, as fire fronts often move too quickly for them to escape.

C Crown bushfires spread through the treetops or 'crowns' of forests. Before long, a wide blanket of forest is fully ablaze.

D High temperatures, low relative humidity and strong winds combine to create high fire danger days.

E What was the flank or side of a bushfire can become the new fire front if there is a wind change.

F Special helicopters can scoop up to 9500 litres of water in 45 seconds and dump the whole lot in just 3 seconds.

G Australia's eucalypt forests not only tolerate fire but also need it in order to survive! The seeds of some eucalypts need the heat of a bushfire to be able to open and grow. The low moisture content of eucalypts means they ignite and burn easily. Their fibrous bark is highly combustible.

H Dry forests provide plenty of fuel. Surface bushfires quickly ignite dry, flammable grass, twigs and branches on the ground.

I By using the wrong building materials, planting eucalypts close to the house and stacking firewood against the house, people can actively contribute to the spread of a bushfire.

J A firebrand is burning fuel that is pushed ahead of the fire front by the wind. Firebrands have been known to travel kilometres from their original source. A spot fire is a new bushfire started by firebrands.

K Properties are more likely to survive if gutters are clear of leaves, lawns and shrubs are trimmed, and there is access to water and hoses. People who defend their house must cover up with cotton or woollen clothing.

4.7.2 Black Saturday

Australia's worst bushfire disaster occurred in Victoria on Saturday 7 February 2009. Australians watched their televisions in disbelief as the scale of the impact became known. One hundred and seventy-three people were killed and many more were badly injured; 2029 properties were destroyed, 7000 people were left homeless, and over 400 000 hectares burned. An estimated one million native animals were also killed. The worst-hit communities were Marysville, Kinglake and Strathewen near Melbourne, but 78 Victorian townships were affected. More than $372 million was raised to aid the victims.

Victoria is located in one of the world's most hazardous bushfire zones. The state has experienced a number of disastrous 'mega-fires'. On

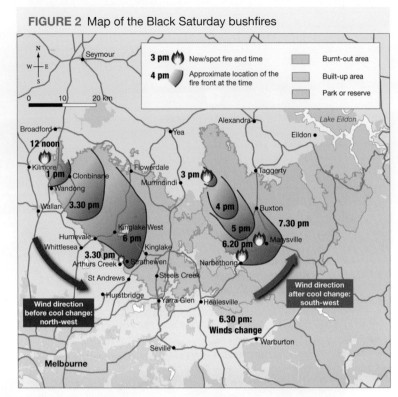

FIGURE 2 Map of the Black Saturday bushfires

Black Thursday, 6 February 1851, a quarter of the new colony of Victoria burned. On 14 February 1926, 60 lives were lost; and on Black Friday, 13 January 1939, another 71 people died. The fires on Ash Wednesday, 16 February 1983, killed 47 Victorians.

These are not the only days when Victoria experienced deadly fires — they're just the worst. Two characteristics mark these historic fire days: a long period of drought and weather conditions on the day that combined high temperatures, low humidity and very strong winds.

Following the 1939 fires, the MacArthur Forest Fire Danger Index (FFDI) was developed. The index uses maximum temperature, relative humidity, wind speed and dryness of fuel (measured using a drought factor) to rate days of fire danger. The ratings are:

Fire danger rating	FFDI range
High	12–25
Very high	25–50
Extreme	>50

The 1939 Black Friday fires scored an FFDI of 100. On Black Saturday, the FFDI for a number of sites in Victoria reached unprecedented levels, ranging from 120 to 190.

Fires are a natural part of the Australian environment, but steps need to be taken to ensure that fires do not have a similar catastrophic impact in future. Governments, planning bodies and individuals need to make decisions about:

- allowing people to build in fire-prone areas
- clearing native vegetation and dead trees around homes and along roadsides
- controlled burning during the cooler months
- provision of private and public bushfire shelters
- early warning systems
- stronger building codes
- deciding whether to stay on one's property and fight the fire or to evacuate.

The impacts of Black Saturday were devastating. Yet we can learn from mistakes that were made to ensure that any future bushfires are managed in a way that minimises potential damage to the environment and to people's property and lives.

FIGURE 3 Anatomy of a bushfire

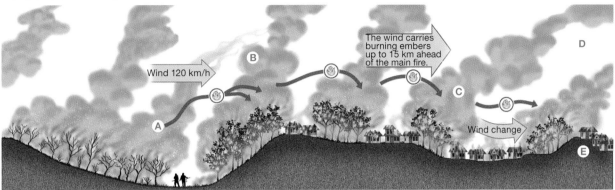

A Fire travels faster uphill, as the hot air rising up the slope preheats the fuel in front of it, and the convection currents produced send burning embers ahead of the fire.

B Smoke column developed into a pyro-cumulus cloud. Lightning started new fires.

C Thick clouds of smoke carry embers and fuel.

D The heat released by the fires was estimated as the energy equivalent of 500 Hiroshima bombs, or enough heat to supply Victoria with electricity for two years.

E More people now live in the outer bush areas around Melbourne, so more people and properties were at risk. Marysville was one of the worst affected towns.

FIGURE 4 A downhill slope will cause a fire to slow down, and an uphill slope will cause a fire to speed up.

4.7 INQUIRY ACTIVITIES

1. How could you make your home and community safer in a bushfire? Think about terrain, climate, vegetation, access to water, firefighting resources and so on. Design a poster outlining one of your ideas. It should be eye-catching and contain a short, clever message. Use the information in this subtopic to help you.

Evaluating, predicting, proposing

2. Write a news report about the scene shown in **FIGURE 1**. Outline the effects on people and wildlife. Include interviews and describe the fire using key terms explained in this topic.   **Describing and explaining**
3. The fires devastated a large area of Victoria. Imagine you are a news reporter. Select a specific location affected by the fires and prepare a three-minute report on some aspect of the fire.

Examining, analysing, interpreting

4.7 EXERCISES

Geographical skills key: GS1 Remembering and understanding **GS2** Describing and explaining **GS3** Comparing and contrasting **GS4** Classifying, organising, constructing **GS5** Examining, analysing, interpreting **GS6** Evaluating, predicting, proposing

4.7 Exercise 1: Check your understanding

1. **GS1** What is a crown bushfire?
2. **GS1** List the firefighting techniques shown in **FIGURE 1**.
3. **GS2** Draw a diagram to explain how eucalypt trees help bushfires spread.
4. **GS2** Why do bushfires often occur in times of drought?
5. **GS1** What started the Black Saturday fires?
6. **GS2** Describe the *changes* that occurred in the direction of the wind on Black Saturday.
7. **GS2** Outline the impacts of this wind change on the fire and the areas affected.
8. **GS2** Describe the *scale* of the impact the Black Saturday fires had.

4.7 Exercise 2: Apply your understanding

1. **GS2** How did the activities of Indigenous Australian peoples affect ecosystems and *environments*?
2. **GS2** What is the FFDI and why is it an important tool for planners and emergency services?
3. **GS2** How did the weather conditions in Victoria on 7 February 2009 contribute to the severity of these fires?
4. **GS6** Imagine a small fire front with a long flank. The fire is being pushed by winds from the north. Suddenly the wind changes and starts blowing from the west. Will the *spaces* and people on the west or on the east of the original fire now be in danger?
5. **GS5** Look carefully at the table below.
   (a) What is relative humidity? Explain how relative humidity might contribute to creating bushfire conditions.
   (b) What combination of weather conditions is most likely to produce bushfires?
   (c) Explain the reasons for your selection.
   (d) If the weather bureau was predicting a top temperature of 40 °C, wind speeds of 75 km/h and humidity of 7 per cent, what would be the fire danger classification?

Fire danger	Temp. (°C)	Relative humidity (%)	Wind speed (km/h)
Low	21	70	10
Moderate	30	20	20
Very high	35	10	50
Explosive	41	5	80

Try these questions in learnON for instant, corrective feedback. Go to www.jacplus.com.au.

4.8 Floods and floodplains

4.8.1 Why does it flood?

Floods are a natural occurrence, but they are a natural hazard to humans, who tend to build farms, towns and transport routes in areas such as floodplains. A floodplain (**FIGURE 1**) is an area of relatively flat land that borders a river and is covered by water during a flood. Floodplains are formed when the water in a river slows down in flat areas. The river begins to meander and gradually deposits **alluvium**, which builds up the floodplain and other landforms such as deltas.

These fertile, flat areas are used for farming and settlement around the world. In Australia, many of our richest farmlands are on floodplains, and towns are often built on them, close to rivers. Such towns are subject to flooding. The possibility of flood is also increased when vegetation in **catchment areas** has been cleared or modified. Native vegetation can slow down run-off and reduce the chance of flooding.

FIGURE 1 Flat, fertile floodplains are often preferred areas for settlement and farming.

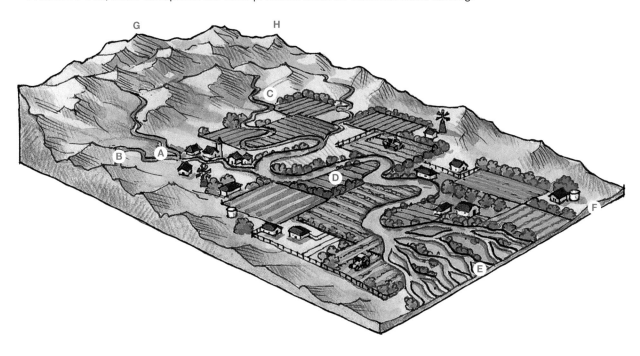

A Tributary stream
B Waterfall
C Meander
D Floodplain

E Delta
F Deposited sediment
G Mountains
H Watershed

4.8.2 La Niña and floods

A La Niña event in Australia is often associated with floods. La Niña is virtually the opposite of El Niño. Very cold waters dominate the eastern Pacific, and the oceans off Australia are warmer than normal. Large areas of low pressure extend over much of Australia; warm, moist air moves in, and above average rainfall occurs. There can also be torrential rain and widespread floods.

Recent La Niña events in Australia occurred in 2010–2011, when many parts of Queensland, New South Wales and Victoria were flooded (**FIGURE 2**).

4.8.3 Types of floods

Even though Australia is the driest of all the world's inhabited continents, there are periods of very heavy rainfall and flood. Flood disasters in Australia damage property, kill livestock and cause the loss of human life. In some cases, entire sections of a town have been washed away, as in 1852, when one-third of the town of Gundagai disappeared.

FIGURE 2 How did the floods affect Rockhampton? Rockhampton before the 2011 flood (a) and (b) after the flood

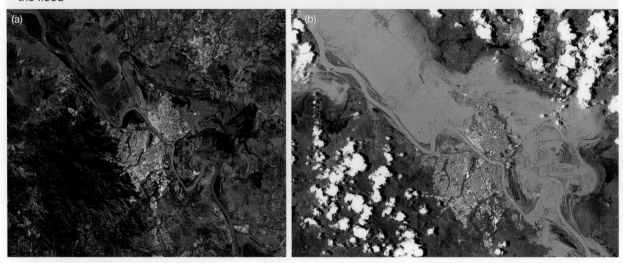

There are three main types of flood:

1. *Slow-onset floods.* These occur along the floodplains of inland rivers, and may last for weeks or months. They are caused by heavy rain and run-off upstream. The water can take days or weeks to affect farms and towns downstream.
2. *Rapid-onset floods.* These occur in mountain headwaters of larger inland rivers or rivers flowing to the coast. The rivers are steeper and the water flows more rapidly. Rapid-onset floods are often more damaging because there is less time to prepare.
3. *Flash floods.* These are caused by heavy rainfall that does not last long, as occurs in a severe thunderstorm. This type of flooding causes the greatest risk to property and human life because it can happen so quickly. It can be a serious problem in urban areas where drainage systems are inadequate.

FIGURE 3 Flooding in Townsville, Queensland, 2019

FIGURE 4 Residents leading a small boat through a flooded street in Townsville, Queensland, 2019

4.8.4 CASE STUDY: The Brisbane floods, 2011

When the Brisbane River broke its banks on 11 January 2011, Australians were shocked and saddened by the devastation left in its wake. Thankfully those affected were able to gain some comfort from the assistance they received from the community as they began the slow process of recovery. However, this is not always an option for those affected by floods in other regions of the world. Australia has experienced several flooding events since 2011, some even more significant than the Brisbane floods. However, a discussion of this event allows us to compare what happened in Brisbane to another flooding event that occurred on the other side of the world on the exact same day.

Queensland, Australia 2010–2011

Country background

Australia is considered a developed nation with a strong economy. Australians earn on average $38 200 per person. Approximately 24 million people live in Australia, with 4.8 million of those living in Queensland. About 84 per cent of all Australians are located within 50 kilometres of the coast.

Why?

The flooding that affected this region was due to a strong La Niña event. Long periods of heavy rain over Queensland catchments caused rivers to burst their banks.

Effects

- Three-quarters of the state was declared a disaster zone.
- At least 70 towns and over 200 000 people were affected.
- There were 35 deaths.
- It cost the Australian economy at least $10 billion.
- Up to 300 roads were closed, including nine major highways.
- Over 20 000 homes were flooded in Brisbane alone.
- There was massive damage and loss of property.

Assistance and recovery

- $1.725 billion was raised by the federal government via a flood levy in the tax system.
- $281.5 million was raised through the Disaster Relief Appeal set up by the then Premier, Anna Bligh.
- Over $20 million was donated to aid agencies such as the St Vincent de Paul Society to help those suffering.
- About $1.2 million was raised through charity sporting events such as Rally for Relief, Legends of Origin and Twenty 20 cricket.
- The Australian Defence Force was mobilised to help with the clean up.
- The Mud Army was formed: 55 000 volunteers registered to help clean up the streets, and thousands more unregistered people joined them.
- Improvements were made to dam manuals to help manage the release of water from dams during floods.

FIGURE 5 Anatomy of a flood

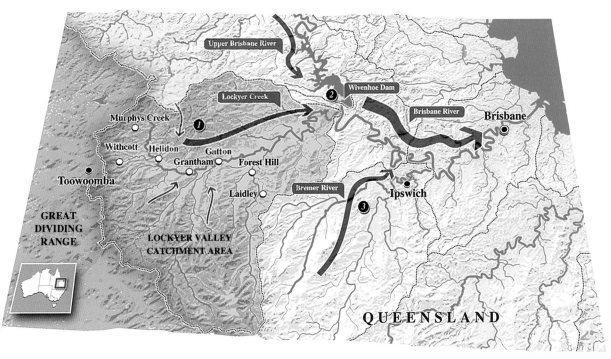

1. Floodwaters from **Lockyer Creek**, which flows into **Brisbane River**. The Lockyer Valley was hit by more than 200 mm of rain.
2. More than 490 000 million litres were released from **Wivenhoe Dam** into Brisbane River.
3. Floodwaters from the **Bremer River**, which is also fed by the **Lockyer Valley**. After passing Ipswich, where it burst its banks, the Bremer River flows into the Brisbane River.

  Town heights above sea level in metres: Toowoomba 700 m Murphys Creek 704 m Withcott 262 m Helidon 143 m Grantham 110 m Gatton 111 m Forest Hill 95 m Laidley 135 m Ipswich 54.8 m Brisbane 28.4 m

4.8.5 CASE STUDY: The Brazil floods, 2011

State of Rio de Janeiro, Brazil, 2011

Country background

Brazil is considered a developing nation. Brazilians earn on average $10 200 per person per year. Approximately 209 million people reside in Brazil, with 650 000 living in the three towns worst affected by the flooding.

Why?

Due to the equivalent of a month's rain falling in 24 hours, flash flooding occurred in a mountainous region in Rio de Janeiro State and São Paulo State. Hillsides and riverbanks collapsed due to landslides. It is believed that illegal construction and deforestation may have contributed to the instability of the land.

Effects

- Approximately 900 people died — most of them in poverty-stricken areas with poor housing conditions and no building policies.
- Forty per cent of the vegetable supply for the city of Rio de Janeiro was destroyed.
- Around 17 000 people were left homeless.
- There was widespread property damage, most of it to homes built riskily at the base of steep hills.

Assistance and recovery

- $460 million was set aside by Brazil's president for emergency aid and reconstruction.
- Troops were deployed to help.
- There were donations of clothes and food to the area from other Brazilians.
- About $450 million was loaned by the World Bank.
- Support was given by internal and international charities.

FIGURE 6 Areas affected by the floods in Brazil, 13 January 2011

Source: Spatial Vision

FIGURE 7 Hills collapsed after the heavy rains, destroying homes.

4.8 INQUIRY ACTIVITIES

1. Create a poster that will warn people about the consequences if they continue to develop on floodplain regions. **Evaluating, predicting, proposing**
2. Imagine you had to evacuate your home because it was under threat from a flood. What five things would you take with you, and why? **Evaluating, predicting, proposing**
3. In a table, classify (in point form) the impacts on people, the economy and the *environment* of the Brisbane and Brazilian floods. **Classifying, organising, constructing**
4. Create an overlay map to show the distribution of water in the aerial photos of Rockhampton before and after the 2011 flood. Estimate the percentage of the area covered in both maps. Describe the *changes* to the *environment*. **Classifying, organising, constructing**

4.8 EXERCISES

Geographical skills key: GS1 Remembering and understanding **GS2** Describing and explaining **GS3** Comparing and contrasting **GS4** Classifying, organising, constructing **GS5** Examining, analysing, interpreting **GS6** Evaluating, predicting, proposing

4.8 Exercise 1: Check your understanding

1. **GS1** What are the three main types of floods?
2. **GS1** What is a floodplain?
3. **GS1** What is a catchment area?
4. **GS2** What is alluvium and why is it important to agriculture?
5. **GS2** Construct a simple flow diagram to show the process of a La Niña event.
6. **GS1** When did the floods in Queensland and Brazil occur?
7. **GS2** Think back on the floods mentioned in this subtopic.
   (a) Explain, in your own words, the three causes of the Brisbane floods.
   (b) Explain the causes of the Brazil floods.
   (c) Compare the causes of these two floods by identifying the similarities and differences between them.
8. **GS3** Read the case studies on the 2011 Brisbane and Brazil floods.
   (a) Compare the *scale* of these floods.
   (b) Give reasons for the differences in the *scale* of these floods.

4.8 Exercise 2: Apply your understanding

1. **GS6** Should people continue to build on floodplains? Why or why not? Think globally when formulating your argument. Consider *environmental*, cultural and economic factors that could have an impact on a person's reasoning when choosing a *place* to settle.
2. **GS2** Why do floods occur on floodplains and in deltas?
3. **GS6** How might the effects of floods in urban *spaces* differ from effects in rural *spaces*?
4. **GS3** How were the responses to the floods in Brisbane and Brazil similar? How and why do you think they were different?
5. **GS3** Suggest a few strategies that could be implemented to lessen the impact of floods if they occurred in Brisbane and Brazil regions again.
6. **GS6** Imagine both Brisbane and Brazil had been given warning that these floods were going to occur. Suggest at least two changes you would expect in relation to the impacts of these two flood events.

Try these questions in learnON for instant, corrective feedback. Go to www.jacplus.com.au.

4.9 Managing the impacts of floods

4.9.1 Flood management

Floods occur in many countries around the world. It is important for these countries to learn how to effectively live with this natural hazard. Managing the effects of floods is important if the amount of damage caused is to be minimised. Unfortunately, not all countries have the same resources to tackle this problem. Those countries that are able to invest in flood-prevention **infrastructure** have a greater chance of reducing the risk of flood.

FIGURE 1 Lake Wivenhoe, Queensland, at 190 per cent capacity, January 2011

The most common form of flood management is to build a barrier that prevents excess water from reaching areas that would suffer major damage. Levees (see **FIGURE 2**), **weirs** and dams are examples of structures that are built to contain floodwaters. Dams that are used to stop flooding need to be kept below a certain level to allow space for floodwater to fill. Wivenhoe Dam in south-east Queensland was built in response to floods in 1974 (see **FIGURE 1**). However, there was some debate about whether this dam could have been used more effectively during the 2010–2011 floods. During these floods, engineers had some difficult decisions to make. If they allowed the dam to become too full, they risked flood waters spilling over the side of the dam and eroding the dam wall. Yet by releasing water through the spillway, the engineers actually contributed to the severity of the flood.

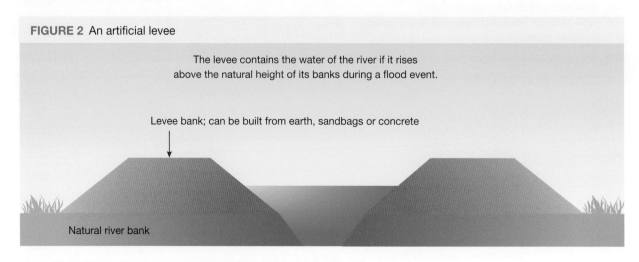

FIGURE 2 An artificial levee

The levee contains the water of the river if it rises above the natural height of its banks during a flood event.

Levee bank; can be built from earth, sandbags or concrete

Natural river bank

To prevent London being flooded during unusually high tides and storm surges, the city constructed the Thames Barrier, a system of floodgates that stretch across the width of the river (see **FIGURE 3**). The barrier is activated when predicted water levels are above a certain height. If this happens, the gates rise to stop the incoming water. Once the water recedes, the danger has passed and the gates are lowered.

FIGURE 3 The Thames flood barrier seen here with its gates up

Another way to manage the risk of damage from floods is to stop building on low-lying land that is guaranteed to flood. Unfortunately, in many urban areas this land has been developed, which increases the chance of property damage in a flood.

Since 2006, the Brisbane City Council has offered a residential property buy-back scheme. This scheme gives people the opportunity to sell their property to the council if they live in a low-lying area that has a 50 per cent chance of flooding every year. People will not be allowed to build on this land again. For this initiative to be successful, it is essential that the price offered by the council is similar to what the owners would get in a private sale; otherwise there is no incentive to use it. The owners of 207 properties were presented with this opportunity yet only 45 had accepted the offer before the 2011 floods swept through Brisbane.

Unfortunately not all countries have the finances to fund property buy-backs or large-scale barrier building. Bangladesh, for example, experiences annual flooding during the **monsoon** season. In response to this, homes are usually built on raised land above flood levels or on stilts.

In order to prepare the population for the arrival of floods, Bangladesh has developed a flood forecasting and warning system that can be broadcast via newspapers, television, radio, the internet and email. Regrettably, due to the growing population in the capital of Dhaka, building is now occurring on low-lying land that was previously used to store floodwater (see **FIGURE 4**). As a result, many people are still being affected by flooding. In 1998, 65 per cent of Bangladesh was inundated. Twenty million people needed shelter and food aid for two months. One thousand and fifty people lost their lives in this flood, which remains the most significant flood in Bangladesh's history.

Around the world floods are becoming more frequent and their impact on people and the environment more damaging and costly. Write an argument that supports and then an argument that would challenge the following viewpoint: People should not be allowed to live in areas that are prone to flooding.

[Critical and Creative Thinking Capability]

FIGURE 4 In Dhaka, homes are built on stilts to avoid the floodwaters.

on Resources

Interactivity	Responding to floods (int-3085)	
Weblink	Bureau of Meteorology	
Google Earth	Lake Wivenhoe	

4.9 INQUIRY ACTIVITY

Use the **Bureau of Meteorology** weblink in the Resources tab to find out more about flood warnings. Prepare an information sheet that could be released to a rural community about to be affected by a major flood event. It should include tips on what to do before, during and after the event. Search for the area you live in and check any flood warnings it has had in the past. **Evaluating, predicting, proposing**

4.9 EXERCISES

Geographical skills key: GS1 Remembering and understanding **GS2** Describing and explaining **GS3** Comparing and contrasting **GS4** Classifying, organising, constructing **GS5** Examining, analysing, interpreting **GS6** Evaluating, predicting, proposing

4.9 Exercise 1: Check your understanding

1. **GS1** What flood management techniques are being used in Brisbane?
2. **GS1** Where are you at most risk from flood when building?
3. **GS2** Look at **FIGURE 2** and use it to help write a definition for the term levee.
4. **GS2** How can an early warning system reduce the risk of a flood disaster?
5. **GS2** Explain the *interconnection* between population growth and the risk posed by floods.

4.9 Exercise 2: Apply your understanding

1. **GS6** What would happen if a dam, built to prevent floods, was already full to capacity and the area received more heavy rainfall? What might be some of the consequences?
2. **GS2** Dams are the most common method used to manage the impacts of floods. Why do you think this is the case?
3. **GS3** The River Thames in London has unique geographic conditions that need to be managed. Explain these conditions and how engineers have attempted to solve the problems these conditions can cause.
4. **GS6** Explain the dilemma which Wivenhoe Dam engineers faced in 2011. What would you have done in the same situation?
5. **GS5** Bangladesh faces severe floods on a yearly basis. Why would building dams not help manage the impact of these floods successfully?

Try these questions in learnON for instant, corrective feedback. Go to www.jacplus.com.au.

4.10 SkillBuilder: Interpreting diagrams

What are diagrams?

A diagram is a graphic representation of something. In Geography, it is often a simple way of showing the arrangement of elements in a landscape and the relationships between those elements. Diagrams also have annotations: labels that explain aspects of the illustration.

Select your learnON format to access:

- an overview of the skill and its application in Geography (Tell me)
- a video and a step-by-step process to explain the skill (Show me)
- an activity and interactivity for you to practise the skill (Let me do it)
- questions to consolidate your understanding of the skill.

Resources

Video eLesson Interpreting diagrams (eles-1636)

Interactivity Interpreting diagrams (int-3132)

4.11 Thinking Big research project: Water access comparison

SCENARIO

The world's water resources are not distributed equally and some countries struggle to find a fresh water supply. You will research and compare two countries that have poor access to fresh water and analyse the natural and human factors involved.

Select your learnON format to access:

- the full project scenario
- details of the project task
- resources to guide your project work
- an assessment rubric.

 Resources

 projectsPLUS Thinking Big research project: Water access comparison (pro-0235)

4.12 Review

4.12.1 Key knowledge summary

Use this dot point summary to review the content covered in this topic.

4.12.2 Reflection

Reflect on your learning using the activities and resources provided.

 Resources

 eWorkbook Reflection (doc-32135)

Crossword (doc-32136)

Interactivity Water variability and its impacts crossword (int-7701)

KEY TERMS

alluvium the loose material brought down by a river and deposited on its bed, or on the floodplain or delta
catchment area the area of land that contributes water to a river and its tributaries
drought a long period of time when rainfall received is below average
economic relating to wealth or the production of resources
El Niño the reversal (every few years) of the more usual direction of winds and surface currents across the Pacific Ocean. This change causes drought in Australia and heavy rain in South America
evaporate to change liquid, such as water, into a vapour (gas) through heat
flood inundation by water, usually when a river overflows its banks and covers surrounding land
incentive something that encourages a person to do something
infrastructure the basic physical and organisational structures and facilities that help a community run, including roads, schools, sewage and phone lines

---

meteorologist a person who studies and predicts weather

monsoon rainy season accompanied by south-westerly summer winds in the Indian subcontinent and South-East Asia

natural disaster an extreme event that is the result of natural processes and causes serious material damage or loss of life

natural hazard an extreme event that is the result of natural processes and has the potential to cause serious material damage and loss of life

precipitation rain, sleet, hail, snow and other forms of water that falls from the sky when water particles in clouds become too heavy

rebate a partial refund on something that has already been paid for

southern oscillation a major air pressure shift between the Asian and east Pacific regions. Its most common extremes are El Niño events.

vulnerability the state of being without protection and open to harm

water vapour water in its gaseous form, formed as a result of evaporation

weir a barrier across a river, similar to a dam, which causes water to pool behind it. Water is still able to flow over the top of the weir.

5 Natural hazards and extreme events

5.1 Overview

A little bit of stormy weather never hurt anyone. But what happens when it becomes extreme?

5.1.1 Introduction

People have long harnessed the power of the wind for energy: we use it to dry clothes, to produce electricity, to pump excess water from the surface of the land and to bring groundwater to the surface.

But strong winds can also cause great destruction, especially when accompanied by heavy rain. These winds can tear roofs from houses and pull trees from the ground.

on Resources

☑ **eWorkbook** Customisable worksheets for this topic

🎞 **Video eLesson** Wind hazards (eles-1618)

LEARNING SEQUENCE

5.1 Overview
5.2 Why does the wind blow?
5.3 **SkillBuilder:** Cardinal points: wind roses `online only`
5.4 Cyclones
5.5 Thunderstorms
5.6 **SkillBuilder:** Creating a simple column or bar graph `online only`
5.7 Typhoons in Asia
5.8 The impact of tornadoes on people and the environment
5.9 When water turns to ice and snow
5.10 Responding to extreme weather events
5.11 **Thinking Big research project:** Weather hazard documentary `online only`
5.12 **Review** `online only`

To access a pre-test and starter questions, and receive immediate, **corrective feedback** and **sample responses** to every question, select your learnON format at www.jacplus.com.au.

5.2 Why does the wind blow?

5.2.1 Air has weight

Earth's atmosphere protects us from the extremes of the sun's heat and the chill of space, making conditions right to support life. The air in the lowest layer of the atmosphere is called the **troposphere**. Weather is the result of changes in this layer of the atmosphere.

The air around us has weight. The weight of the air above us pushes down on the surface, creating pressure. If we could tie a **barometer** to a hot air balloon, we would see the pressure readings fall as the balloon rose in the atmosphere. This is because there is less air higher up in the atmosphere. You may have read about mountain climbers and athletes having difficulty breathing when they are at high altitudes.

Air pressure

When a person blows up a balloon, the pressure inside the balloon is higher than the surrounding air. When the neck of the balloon is released, the air rushes out of the balloon, as shown in **FIGURE 1 (a)** and **(b)**. This is wind. If we did not have wind, temperatures would continue to rise over the equator and decrease at the poles.

FIGURE 1(a) The pressure inside a balloon is higher than the surrounding air.

FIGURE 1(b) When the neck of the balloon is released, air rushes out, moving from a space of high pressure to one of low pressure.

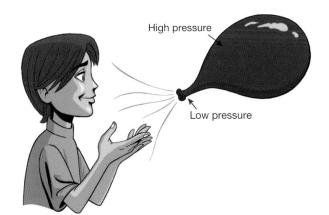

High pressure

Low pressure

Meteorologists are able to measure air pressure using a unit of measure called a millibar. The average weight of air is about 1013 millibars. Measurements higher than this indicate areas of high pressure; here, the air is sinking. Measurements lower than 1013 millibars indicate areas of low pressure; here, the air is rising. Wind is caused by air moving from areas of high pressure to areas of low pressure.

5.2.2 Why does air pressure vary across the Earth?

Variations in air pressure are the result of the heating effect of the sun and the rotation of the Earth.

Effects of the sun

The warming influence of the sun varies with the time of day (see **FIGURE 2**) and latitude (distance from the equator). Temperatures are higher in the middle of the day, and higher at the equator than at the poles. Warm air is also less dense than cold air. This is because as the air heats, it expands, causing it to rise. Air pressure over the equator is less than at the poles. As the warm air over the equator rises and expands, cooler air from near the poles rushes in to replace it. As a result, air is circulated around the Earth, and this movement of air is what we call wind.

FIGURE 2 On a smaller scale, this diagram shows the effect of the sun on a sea breeze.

Land heats up and cools down more quickly than the sea.

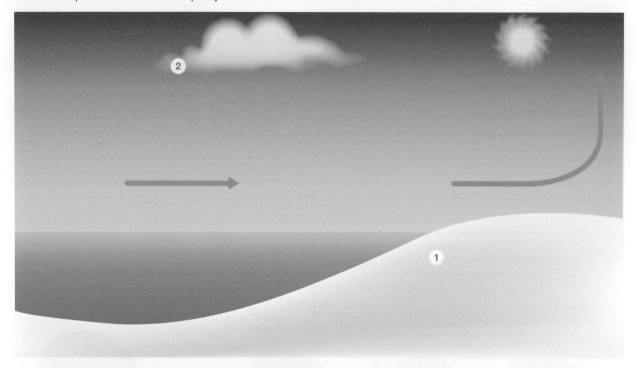

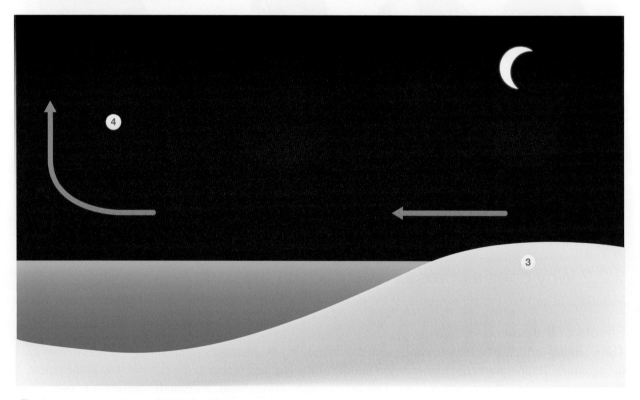

1. During the day the air above the land heats up, expands and rises.
2. The atmospheric pressure above the land drops and air moves in from above the sea, where the air pressure is higher. This causes a sea breeze or an onshore breeze.
3. During the evening, the temperature of the land drops much faster than the temperature of the sea.
4. The air above the sea becomes hotter than the air above the land, so it rises and a breeze flows from the coast out to sea, reversing the effect.

Effect of the Earth's rotation

The rotation of the Earth on its axis causes the air above the surface of the Earth to be deflected rather than to travel in a straight line. This causes the wind to circle around high- and low-pressure systems. The direction in which winds circle depends on whether you are in the northern or southern hemisphere. As the air moves from an area of high pressure to an area of low pressure, winds circle in the opposite direction in each hemisphere. In an area of high pressure, the winds circle in an anticlockwise direction in the southern hemisphere and a clockwise direction in the northern hemisphere. This deflection of winds is known as the Coriolis effect (see **FIGURE 3**).

FIGURE 3 Wind is caused by air moving from areas of high pressure to areas of low pressure. Its direction is influenced by the rotation of the Earth.

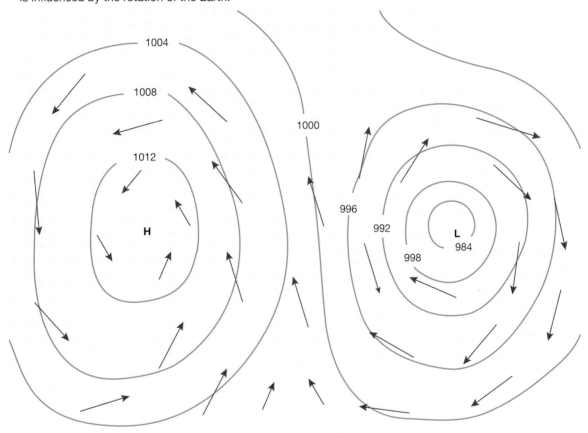

5.2.3 How is wind shown on a weather map?

Differences in air pressure lead to variations in the strength of the wind. You can work out the strength of the wind by looking at weather maps, the behaviour of objects or by using instruments designed to measure the strength of the wind. Winds are named according to their source. This means that a northerly wind is coming from the north and a southerly from the south.

If you study the **isobars** on a weather map you will notice that they are not evenly spaced. Look closely at the map in **FIGURE 4**. The wind is strongest in the southern regions of this map, where the isobars are close together, and gentler in the northern parts of the map, where the spacing between them is much greater.

FIGURE 4 A typical weather map

Greater distance between isobars indicates wind is weaker in these regions.

Cold front advancing from west to east — towards southern Australia. Note how the isobars behind it are close together, indicating strong winds.

Shading indicates rain fell in the previous 24 hours.

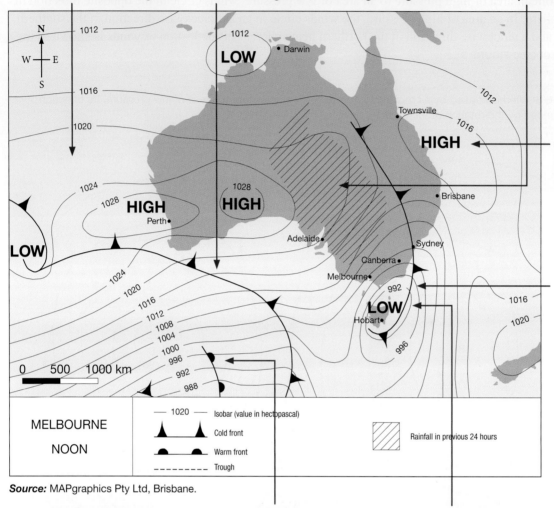

High pressure system. Note how the isobar is joining points with an atmospheric pressure of 1016 (higher than 1013 average).

The cold front that has just passed over Melbourne has brought unsettled, rainy weather behind it. More rainy weather is likely for Melbourne when the cold front coming towards it from the west gets closer.

Source: MAPgraphics Pty Ltd, Brisbane.

The main influences on Melbourne's weather are the low pressure system south of it, the high pressure system to the north-west, and the cold fronts to the west and east.

Warm front advancing from west to east behind the cold front. Note the lower atmospheric readings on the isobars.

Low pressure system. Note how the isobar is joining points with an atmospheric pressure of 992 (lower than the 1013 average).

The symbols shown in **FIGURE 5** are also commonly used on weather maps to give a more accurate representation of wind speed and to provide information on the direction of the wind.

FIGURE 5 Symbols commonly used to indicate wind strength

○	calm (0–2 km/h)	3–7 km/h		8–12 km/h
	13–17 km/h	18–22 km/h		23–27 km/h
	28–32 km/h	33–37 km/h		38–42 km/h
	43–47 km/h	48–52 km/h		53–57 km/h

FIGURE 6 How to read wind symbols on a weather map

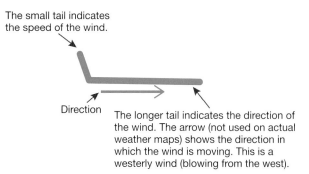

The small tail indicates the speed of the wind.

Direction

The longer tail indicates the direction of the wind. The arrow (not used on actual weather maps) shows the direction in which the wind is moving. This is a westerly wind (blowing from the west).

Can I make observations about the speed of the wind?

The Beaufort scale (see **FIGURE 7**) relates wind speed to the observable movement of objects within the environment.

FIGURE 7 The Beaufort scale is based on the observable impact of winds.

0 Calm
Less than 2 km/h
Smoke rises vertically

2 Light breeze
6–12 km/h
Wind felt on face, wind vanes move

4 Moderate breeze
21–30 km/h
Dust and loose paper move, small branches move

6 Strong breeze
41–51 km/h
Large branches move, umbrellas difficult to use, difficult to walk steadily

8 Gale
64–77 km/h
Twigs broken off trees, difficult to walk

10 Whole gale
88–101 km/h
Trees uprooted, considerable structural damage

12 Hurricane/cyclone
Greater than 120 km/h
Widespread devastation

1 Light air
2–5 km/h
Smoke drift shows wind direction, wind vanes don't move

3 Gentle breeze
13–20 km/h
Leaves and small twigs in motion, hair disturbed, clothing flaps

5 Fresh breeze
31–40 km/h
Small trees with leaves begin to sway, wind force felt on body

7 Moderate gale
52–63 km/h
Whole trees in motion, inconvenience felt when walking

9 Strong gale
78–86 km/h
People blown over, slight structural damage, including tiles blown off houses

11 Storm
102–120 km/h
Widespread damage

Using a wind rose

A wind rose such as that shown in **FIGURE 8** uses data collected over long periods of time to visually represent wind information. The spokes represent wind direction; the longer the spoke the more frequently the wind blows from a particular direction. The thickness of the bands represents the speed of the wind. Refer to the SkillBuilder 'Cardinal points: wind roses' in subtopic 5.3 to learn how to use a wind rose.

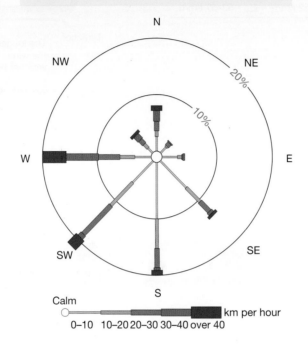

FIGURE 8 A wind rose can show wind speed, direction and frequency over a long period of time.

Calm — km per hour
0–10 10–20 20–30 30–40 over 40

on Resources

Interactivity Highs and lows (int-3086)

Weblink Wind rose maps

Explore more with myWorldAtlas

Deepen your understanding of this topic with related case studies and questions.
- Exploring places > Australia > **Australia: weather and climate**
- Building skills > Understanding direction > **Wind and sun direction**

5.2 INQUIRY ACTIVITIES

1. Working with a partner, try the following activity to illustrate the influence of the Earth's rotation on wind.
   - *Step 1.* Place a pin through the centre of a piece of paper and attach it to a piece of cardboard. Make sure the paper can move freely on the pin.
   - *Step 2.* Have your partner rotate the piece of paper around the pin. At the same time, you should attempt to draw a straight line on the page.
   - *Step 3.* Record your findings.
   - *Step 4.* Compare and discuss your results with the class.

Classifying, organising, constructing

2. Over the course of the next week, collect weather maps from the daily newspaper and find your location.
   (a) Is the weather being influenced by a high- or a low-pressure system?
   (b) Will the wind be moving in a clockwise or anticlockwise direction? Give reasons for your answer.
   Examining, analysing, interpreting
3. How easy is it to predict the weather? People often complain that the forecasters don't always get it right.
   (a) Using the weather maps you collected earlier, and the observations you made, write a weather forecast for tomorrow. In your forecast, refer to both wind speed and direction.
   (b) Collect tomorrow's weather map and make observations. Record your findings.
   (c) Compare what you have written for this activity. How accurate were your predictions? Suggest factors that might influence the accuracy of such predictions and *changes* that you observe.
   Comparing and contrasting
4. Collect a weather map from the newspaper or online. Based on what you know about weather and reading weather maps, predict what the wind conditions will be like in each of the Australian capital cities over the next day or so. Use the **Wind rose maps** weblink in the Resources tab to compare your predictions with what is shown on these maps. Make sure you select the current month. Note any similarities and differences. Why might differences occur?
   Evaluating, predicting, proposing
5. Devise your own symbols to indicate wind strengths on the Beaufort scale. You could use the common weather map symbols shown in **FIGURE 5** as a starting point. Obtain a current weather map from the newspaper or online. Paste it onto a sheet of paper and annotate your map with your symbols for describing wind speed. Swap maps with a partner and further annotate each other's maps with written descriptions of the symbols shown.
   Classifying, organising, constructing

5.2 EXERCISES

Geographical skills key: GS1 Remembering and understanding **GS2** Describing and explaining **GS3** Comparing and contrasting **GS4** Classifying, organising, constructing **GS5** Examining, analysing, interpreting **GS6** Evaluating, predicting, proposing

5.2 Exercise 1: Check your understanding

1. **GS1** What is wind?
2. **GS2** Identify the two factors that influence wind. Would either of these factors influence the strength of the wind? Explain.
3. **GS2** Explain why you are not affected by the pressure of the atmosphere.
4. **GS1** What role does the sun play in causing wind?
5. **GS1** What *change* does difference in air pressure cause?
6. **GS2** Describe two methods you could use to determine wind speed.
7. **GS6** In your opinion, which of these methods gives the most useful information about wind speed? Give reasons for your answer.

5.2 Exercise 2: Apply your understanding

1. **GS3** What is the *interconnection* between our atmosphere and the weather we experience at the Earth's surface?
2. **GS5** Using **FIGURE 4**, describe the wind speeds and directions in Western Australia and along the east coast of Australia on that day.
3. **GS4** Refer to **FIGURES 5** and **6**. Draw wind symbols to represent the following wind speeds and directions:
   (a) northerly; 28–32 kph
   (b) south-westerly; 8–12 kph
   (c) easterly; 13–20 kph.
4. **GS6** Refer to **FIGURE 4**; a cold front is approaching Brisbane. Describe how the weather will *change* when the cold front arrives.
5. **GS3** Explain the relationship between the Coriolis effect and wind.

Try these questions in learnON for instant, corrective feedback. Go to www.jacplus.com.au.

5.3 SkillBuilder: Cardinal points: wind roses

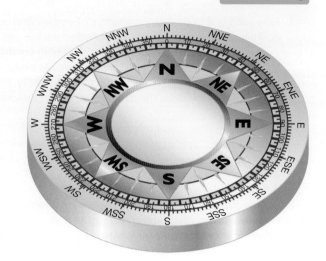

What are wind roses?

A wind rose is a diagram that shows the main wind features of a place, in particular, wind direction, speed and frequency. Wind directions can be divided into 8 or 16 compass directions.

Select your learnON format to access:

- an overview of the skill and its application in Geography (Tell me)
- a video and a step-by-step process to explain the skill (Show me)
- an activity and interactivity for you to practise the skill (Let me do it)
- questions to consolidate your understanding of the skill.

 Resources

 Video eLesson   Cardinal points: wind rose (eles-1638)

 Interactivity   Cardinal points: wind rose (int-3134)

5.4 Cyclones

5.4.1 What is a cyclone?

Tropical **cyclones** (called hurricanes in the Americas and **typhoons** in Asia) can cause great damage to property and significant loss of life. Some 80 to 100 tropical cyclones occur around the world every year in tropical coastal areas located north and south of the equator. Australia experiences, on average, about 13 cyclones per year.

Cyclones form when a cold air mass meets a warm, moist air mass lying over a tropical ocean with a surface temperature greater than 27 °C. Cold air currents race in to replace rapidly rising, warm, moist air currents, creating an intense low-pressure system. Winds with speeds over 119 kilometres per hour can be generated. Cyclones are classified using the scale in **TABLE 1**.

TABLE 1 Classification of cyclones using the Saffir–Simpson scale

Category	Wind gust speed/ocean swell	Damage
1	Less than 125 km/h 1.2–1.6 m	Mild damage
2	126–169 km/h 1.7–2.5	Significant damage to trees
3	170–224 km/h 2.6–3.7 m	Structural damage, power failures likely
4	225–279 km/h 3.8–5.4 m	Most roofing lost
5	More than 280 km/h More than 5.4 m	Almost total destruction

---

FIGURE 1 World distribution of tropical cyclones by names used in different regions

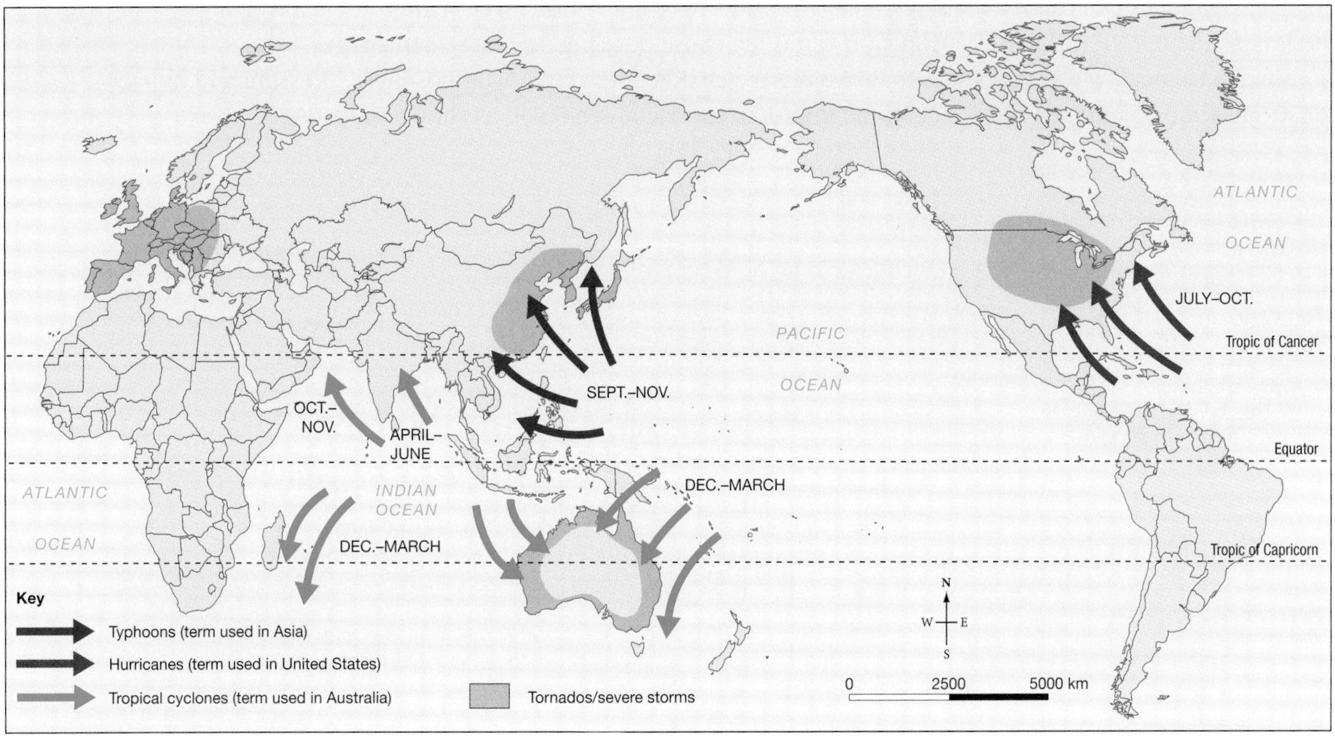

ATLANTIC
OCEAN
PACIFIC
OCEAN
INDIAN
OCEAN
ATLANTIC
OCEAN
JULY–OCT.
Tropic of Cancer
SEPT.–NOV.
OCT.–
NOV.
APRIL–
JUNE
Equator
DEC.–MARCH
DEC.–MARCH
Tropic of Capricorn
Key
Typhoons (term used in Asia)
Hurricanes (term used in United States)
Tropical cyclones (term used in Australia)
Tornados/severe storms
N
W
E
S
0
2500
5000 km

Source: Spatial Vision

FIGURE 2 shows the continuous cycle of evaporation, condensation and precipitation associated with cyclones. At first the winds spin around an area about 200 to 300 kilometres wide. As the winds gather energy by sucking in more warm moist air, they get faster. In severe cyclones, winds may reach speeds of 295 kilometres per hour. The faster the winds blow, the smaller the area around which they spin; this is called the eye. It might end up being only about 30 kilometres wide. Around the edge of the eye, winds and rain are at their fiercest. However, in the eye itself, the air is relatively still, and the sky above it may be cloudless.

FIGURE 2 How a cyclone forms

5
6
4
7
3
8
1
2

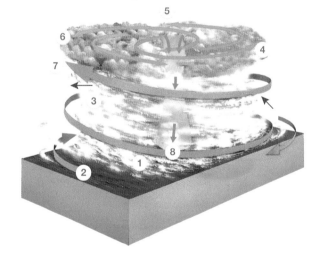

1 Warm sea water evaporates and rises.
2 Low pressure centre creates converging winds, which replace rising air.
3 Warm air spirals up quickly.
4 Warm moist air is drawn in, providing additional energy.
5 Water vapour fuels cumulus clouds.
6 In the upper atmosphere, the air moves away from the eye.
7 Storm moves in direction of prevailing wind.
8 Descending air in the eye of cyclone

What damage is caused by tropical cyclones?

Tropical cyclones can cause extensive damage if they cross land. **Gale force winds** can tear roofs off buildings and uproot trees. **Torrential rain** can often cause flooding, as can **storm surges**.

When a tropical cyclone approaches or crosses a coastline, the very low atmospheric pressure and impact of strong winds on the sea surface combine to produce a rise in sea level, as shown in **FIGURE 3**.

FIGURE 3 Flooding caused by storm surges

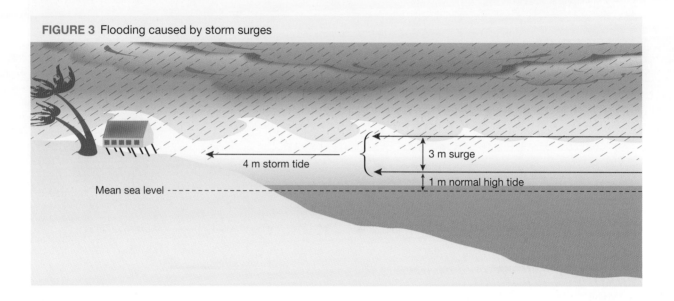

4 m storm tide

3 m surge

1 m normal high tide

Mean sea level

FIGURE 4 Satellite image of Hurricane Katrina, which caused massive damage in New Orleans, United States, in 2005

5.4.2 CASE STUDY: How did Cyclone Winston affect Fiji?

On 7 February 2016, a tropical disturbance was noted north-west of Port Vila, Vanuatu, tracking in a south-easterly direction. By 11 February it had acquired gale-force winds. Over the next few days Cyclone Winston went through a cycle of intensifying, weakening and stalling until finally developing into a category 5 cyclone on 19 February. The following day, shortly before making landfall on Viti Levu, Fiji, Cyclone Winston reached its peak intensity. Sustained winds of 230 kilometres per hour, with gusts of up to 285 kilometres per hour and a central pressure reading of 915 millibars, were recorded.

FIGURE 5 A track map of Cyclone Winston

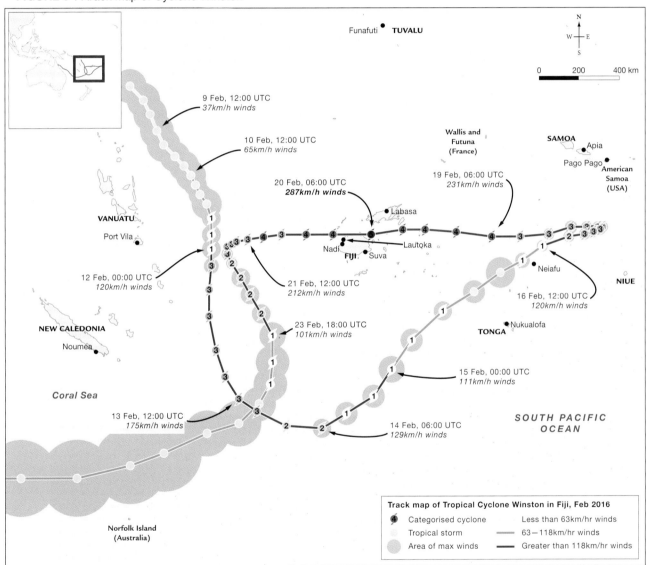

Source: National Hurricane Center, National Oceanic and Atmospheric Administration

FIGURE 6 The power of a cyclone

How much damage was caused?

Cyclone Winston has been described as the most powerful storm to strike in the southern hemisphere. Strong winds battered the island nation of Fiji with damage multiplied by a 4-metre storm surge.

The damage bill has been estimated at more than US$650 million. More than 40 people were killed and communication was cut, leaving at least six outer islands isolated for days. In the years prior to Cyclone Winston, the Fijian government had invested heavily in infrastructure, much of which was washed away. Homes and community facilities were flattened in many communities, with some villagers losing all their possessions (see **FIGURES 7** and **8**). Large regions were left without electricity and water. A week after the cyclone around 45 000 people were still living in evacuation centres.

Fiji's largest industries are sugar cane and tourism. The sugar cane industry alone suffered around US$83 million worth of loss. This figure does not take into account the more than 200 000 people who depend on this industry for their livelihood. Additionally, thousands of acres of root crops were lost.

FIGURE 7 Whole communities were left devastated.

FIGURE 8 Additional damage was caused by a 4-metre storm surge.

The damage to the tourism industry was mixed. While Denaru Island resorts were still able to operate, this was not the case on some of the outer islands. Despite their losses, many of the local villages that depend on tourism were encouraging tourists to return and they were still operating.

What aid has Australia provided?

Both Australia and New Zealand were quick to aid Fiji. Australia worked not only with the Fijian government, but also with the island nation of Tonga, which was also affected by Cyclone Winston (see **FIGURE 9**).

FIGURE 9 Australia's aid operation

The Australian Government has provided

$15 million in response to #TCWinston in Fiji

Australian support will reach up to **200,000 people.**
This includes humanitarian emergency supplies for over
100,000 people provided through **Red Cross, UN
agencies and NGOs** and delivered by the ADF

 Our assistance will help to **restore education and
health services** and support livelihoods for those
affected by the cyclone.

Australian Medical Assistance Teams are providing
lifesaving healthcare in affected communities

ADF is providing:

HMAS Canberra
deployed with 60 tonnes
of emergency supplies
and personnel to repair
critical infrastructure

7 MRH-90 Helicopters
delivering personnel and
essential supplies to remote
localities

Airlifts
from Australia
delivering
humanitarian
support

Follow @AusHumanitarian on Twitter for updates on Australia's humanitarian response in Fiji

UPDATED 9 Mar 2016

Were other areas affected?

The east coast of Australia experienced large waves in the wake of Cyclone Winston, forcing the closure of some popular tourist beaches. Despite warnings from authorities, surfers risked serious injury and even death to take advantage of the huge swells created along the New South Wales and Queensland coastlines. Beaches were still closed a week after Fiji was devastated.

5.4.3 CASE STUDY: What impact did Cyclone Marcia have?

On 15 February 2015, a tropical low developed as a low-pressure system over the Coral Sea. The system drifted in an easterly direction before intensifying into a category 1 cyclone on 18 February. Named Marcia by the Australian Bureau of Meteorology, this cyclone turned towards the south-west and continued to **intensify**, reaching category 2 later that day.

Over the next 12 hours Cyclone Marcia intensified rapidly, reaching category 4 before making a sharp turn towards the south. By 4am on 20 February, Cyclone Marcia was classified as a category 5 system. This rapid rate of development far exceeds the average rate of intensification worldwide.

Indications of the true intensity of Cyclone Marcia, as a category 5 system, have been estimated from satellite and radar imagery. This is because the automatic weather station at Middle Percy Island, where Marcia first made landfall, was located to the west of Marcia's core and outside the eye wall where wind speeds are at their maximum. As a result, only wind speeds equivalent to a category 3 cyclone were recorded (see **FIGURE 10**).

FIGURE 10 The track and intensity of Cyclone Marcia

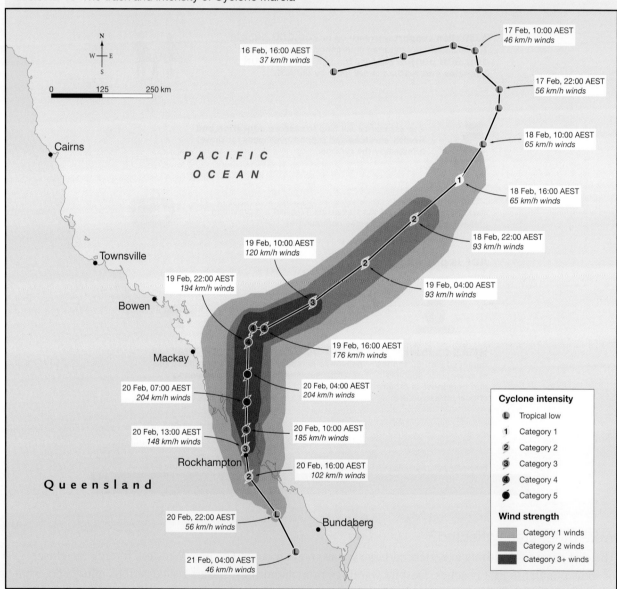

Source: Bureau of Meteorology

Cyclone Marcia was a relatively compact system when compared to other systems, such as Tropical Cyclone Yasi, which caused widespread damage as it carved a 400-kilometre path of destruction. Because Cyclone Marcia was a compact system it weakened quickly after crossing the coast and was far less destructive than similar systems, and only affected places on its track path.

Marcia was downgraded to a tropical low on 21 February, as it tracked to the south of Monto before turning to the south-west where it crossed the Sunshine Coast and moved back out to sea.

Most of the areas affected by Marcia were largely uninhabited; however, the townships and surrounding areas of Byfield, Yeppoon and Rockhampton sustained significant damage. Accurate wind speeds are not available because the strongest part of the eye wall missed the automatic weather stations; therefore, recorded wind speeds were equivalent to a category 3 system.

The damage bill for Cyclone Marcia is expected to reach $750 million. By contrast, Cyclone Yasi left a damage bill of $1.5 billion when it struck in 2011.

Storm surges

Storm surges ranging from 0.5 metres to 2 metres were recorded along the Capricorn Coast. However, as these storm surges coincided with a falling tide there were no significant reports of damage as a result of inundation. Significant beach erosion was evident as sand was stripped from beaches and large numbers of pumice stones were deposited along beaches north of Yeppoon.

Wind damage

Much of the area to the north of Rockhampton is uninhabited; a military training base is located in the area. Most damage was in the Rockhampton region — 97 percent of the region was left without power.

Cyclone Marcia had:
- brought down 2000 power lines
- left 65 000 residents without power
- severely damaged hundreds of homes
- defoliated, snapped in half or uprooted tens of thousands of trees
- inflicted $90 million damage on the agricultural sector, with the banana industry heavily affected.

FIGURE 11 Rockhampton underwater

FIGURE 12 Coastal defoliation and erosion

FIGURE 13 Structural damage in Yeppoon

FIGURE 14 Trees were uprooted by Cyclone Marcia.

5.4 INQUIRY ACTIVITY

Create a timeline showing the development of Cyclone Marcia, from formation to dissipation.

Classifying, organising, constructing

5.4 EXERCISES

Geographical skills key: GS1 Remembering and understanding **GS2** Describing and explaining **GS3** Comparing and contrasting **GS4** Classifying, organising, constructing **GS5** Examining, analysing, interpreting **GS6** Evaluating, predicting, proposing

5.4 Exercise 1: Check your understanding

1. **GS1** What conditions do tropical cyclones need in order to develop?
2. **GS1** What names are given to tropical cyclones in other *places*?
3. **GS1** Why do tropical cyclones die out if they move inland?
4. **GS2** Explain the *changes* that a storm surge can cause to a coastal area.
5. **GS2** How does the *scale* of a cyclone vary?
6. **GS2** What is the *interconnection* between the warmth of seawater and cyclones?
7. **GS5** Study the track map of Cyclone Winston (**FIGURE 5**). Use this map to describe the *scale* of the cyclone and the damage that was caused.
8. **GS2** Explain why Cyclone Winston had an impact on more than one *place*.
9. **GS6** Which phase of Cyclone Marcia would have been the most dangerous? Give reasons for your answer.
10. **GS2** Refer to the case study on Cyclone Marcia (section 5.4.3) and complete the following questions.
    (a) Explain how the extent of damage caused by the storm surge could have been greater if it had occurred at a different time of day.
    (b) Explain what conditions would have been like during the cyclone.
    (c) Describe the damage caused by Cyclone Marcia

5.4 Exercise 2: Apply your understanding

1. **GS6** Describe how the damage would differ between a category 1 and category 5 cyclone.
2. **GS6** Suggest why people are more likely to be killed or injured after the eye of the cyclone has passed.
3. **GS6** Cyclones are associated with destructive winds and the displacement of large volumes of water. Which of these events do you think would cause the most damage to the natural and built *environment*? Justify your answer.
4. **GS5** Refer to **FIGURE 1**, which shows the world pattern of tropical cyclones over *space*.
    (a) When do most cyclones occur north of the equator? When do most cyclones occur south of the equator? Suggest a reason for this difference.
    (b) Name the parts of Australia most at risk from cyclone activity.
5. **GS2** If the water source for cyclones is the ocean over which they form, explain why strong winds and flooding occur in *places* inland from the coast.
6. **GS6** Why could we consider that tropical cyclones are an example of the water cycle at work? Give reasons for your answer.
7. **GS6** Why do you think Cyclone Winston was able to develop into a much stronger storm rather than dissipate once it had affected Tonga?
8. **GS2** Explain the *interconnection* between Cyclone Winston and large waves that resulted in Australian beaches being closed in Queensland and New South Wales.
9. **GS2** Explain why there was not significant loss of life during Cyclone Marcia.
10. **GS6** Suggest reasons why the overall damage bill for Cyclone Marcia was significantly less than that for Cyclone Yasi.

Try these questions in learnON for instant, corrective feedback. Go to www.jacplus.com.au.

5.5 Thunderstorms

5.5.1 What causes thunderstorms?

Thunderstorms, also referred to as electrical storms, form in unstable, moist atmospheres where powerful updrafts occur, which happens when a cold front approaches. It is estimated that, around the Earth, there are 1800 thunderstorms each day. Between 2000 and 2018, an average of around 100 severe thunderstorms were reported in Australia each year.

Some 1000 or so years ago, the Vikings thought thunder was the rumble of Thor's chariot. (He was their god of thunder and lightning.) Lightning marked the path of his mighty hammer Mjöllnir when he threw it across the sky at his enemies.

Today we know that thunderstorms occur when large **cumulonimbus clouds** build up enough static electricity to produce lightning, as shown in **FIGURE 1**. Lightning instantly heats the air through which it travels to about 20 000 °C — more than three times as hot as the surface of the sun. This causes the air to expand so quickly that it produces an explosion (thunder). The time between a lightning flash and the crash of thunder tells you how far away the lightning is (5 seconds means that the lightning is 1.6 kilometres away).

FIGURE 1 How a thunderstorm works

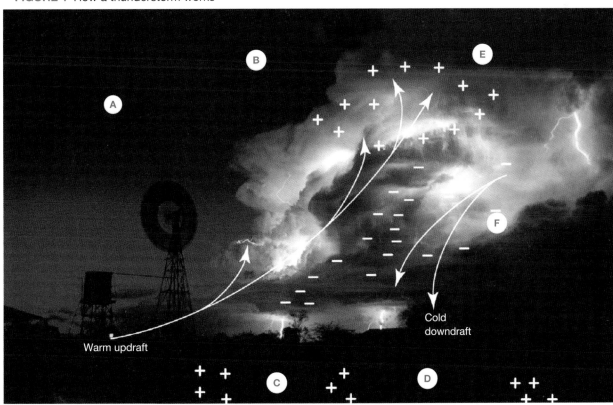

A   As air currents in a cumulonimbus cloud become more violent, they fling ice crystals and water droplets around faster. The more these crystals and droplets smash into one another, the more friction builds up. This creates huge energy stores of static electricity in the cloud.

B   Lighter particles with a positive electric charge drift upwards. Heavier particles with a negative charge sink.

C   The ground below the cloud has a positive charge.

D   Lightning travels to the ground via the shortest route. This is why it sometimes strikes buildings or tall trees.

E   A bolt of lightning actually consists of a number of flashes that travel up and down between the cloud and the ground. This happens so quickly we can't see it.

F   The difference in energy between the positive charge on the ground and the massive negative charge at the bottom of the cloud becomes huge. A lightning bolt corrects some of this difference.

5.5.2 Severe thunderstorms

According to the Bureau of Meteorology, a thunderstorm can be classified as severe if it has one or more of the following features.

- Flash flooding. Thunderstorms often move slowly, dropping a lot of precipitation in one area. The rain or hail may consequently be too heavy and long-lasting for the ground to absorb the moisture. The water then runs off the surface, quickly flooding local areas.
- **Hailstones** that are two centimetres or more in diameter. The largest recorded hailstone had a circumference of 47 centimetres.
- Wind gusts of 90 kilometres per hour or more. Cold blasts of wind hurtle out of thunderclouds, dragged down by falling rain or hail. When the drafts hit the ground, they gust outwards in all directions.

In the right conditions, tornadoes can occur (see subtopic 5.8). These are rapidly spinning updrafts of air that can develop as a result of thunderstorm activity. Although severe tornadoes are not common in Australia, around 400 tornadoes have been recorded here.

FIGURE 2 The Vikings believed the god Thor produced lightning and thunder.

5.5.3 When do thunderstorms occur?

Thunderstorms can occur at any time of the year, but they are more likely to occur during spring and summer, as shown in **FIGURES 3** and **4**. This is due mainly to the warming effects of the sun and the fact that warm air can hold more moisture than cold air.

Thunderstorms are created when cooler air begins to push warmer, humid air upwards. As the warm air continues to rise rapidly in an unstable atmosphere, the cloud builds up higher and begins to spread. Thunderstorms can quickly develop when the atmosphere remains unstable or when it is able to gather additional energy from surrounding winds.

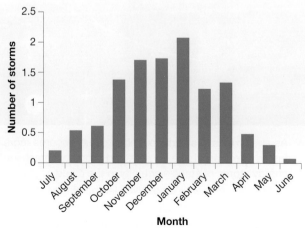

FIGURE 3 Average monthly distribution of thunderstorms in Darwin

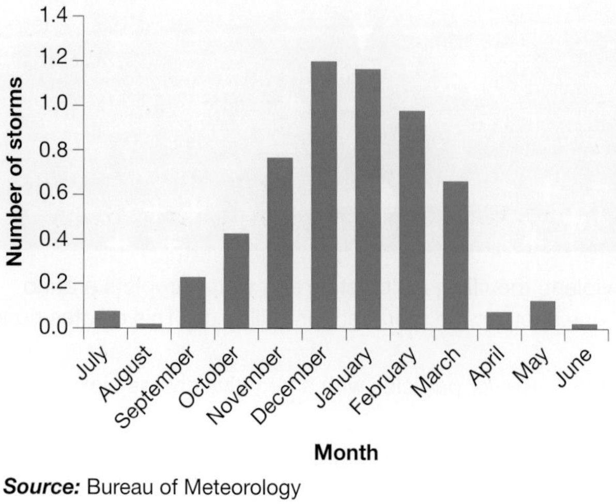

FIGURE 4 Average monthly distribution of thunderstorms in Melbourne

Source: Bureau of Meteorology

Source: Bureau of Meteorology

The time of day when thunderstorms are more likely is shown in **FIGURE 5**. You will notice that thunderstorm activity is greater in the afternoon. This is linked to the daily heating of the Earth by the sun, which peaks in the afternoon.

FIGURE 5 Hourly distribution of thunderstorms in New South Wales and the Australian Capital Territory

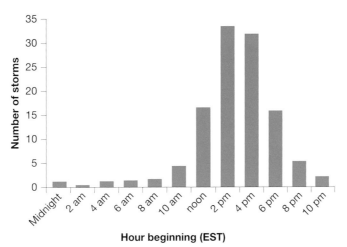

Source: Bureau of Meteorology

5.5.4 What is a hailstorm?

When we think about thunderstorms, we often think only of the high winds, thunder and lightning, but significant damage is also caused by hailstones. Any thunderstorm that produces hailstones large enough to reach the ground is known as a **hailstorm**. Hailstones in Australia tend to range in size from a few millimetres to the size of a tennis ball (see **FIGURE 6**)

FIGURE 6 Hailstones can be the size of a golf ball or even bigger.

5.5.5 Inside a storm

January 2016 saw widespread supercell storm activity across Queensland, New South Wales, Victoria and South Australia.

On 13 January, Melbourne sweltered through temperatures of about 43 °C. Intense thunderstorm activity with wind gusts up to 100 kilometres per hour swept through in the early evening causing the city to be blanketed by a cloud of dust. Up to 1000 homes were left without power.

The following day a severe storm struck Sydney with winds gusting up to 98 kilometres per hour bringing down power lines, damaging buildings and cars and causing flash flooding. More than 40 000 homes and businesses reported power outages. The temperature plummeted by more than 10 °C in five minutes. Emergency services responded to 145 storm-related incidents, including a gas leak.

On 16 January, Townsville recorded 91 millimetres of rain in 30 minutes, resulting in flash flooding that left many motorists stranded. The rain continued to fall, with 181 millimetres recorded in two hours. Wind gusts of more than 100 kilometres accompanied the massive storm that has been described as a once-in-a-100-year event. Unfortunately, while large areas were inundated, the rain had little impact on the region's water storages.

FIGURE 7 How a supercell hailstorm forms

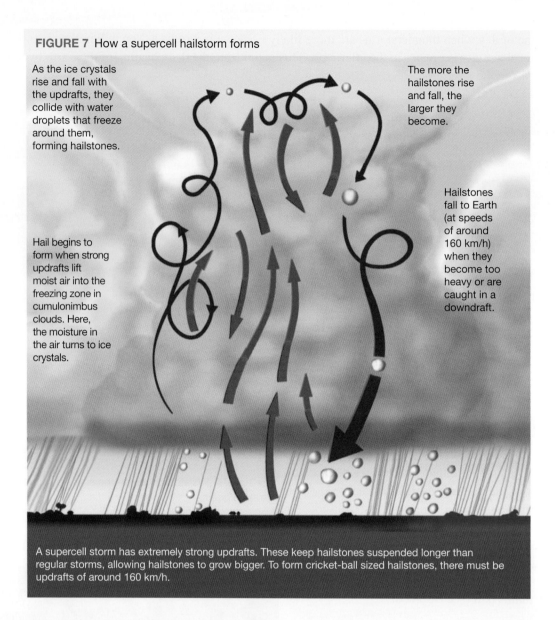

As the ice crystals rise and fall with the updrafts, they collide with water droplets that freeze around them, forming hailstones.

The more the hailstones rise and fall, the larger they become.

Hail begins to form when strong updrafts lift moist air into the freezing zone in cumulonimbus clouds. Here, the moisture in the air turns to ice crystals.

Hailstones fall to Earth (at speeds of around 160 km/h) when they become too heavy or are caught in a downdraft.

A supercell storm has extremely strong updrafts. These keep hailstones suspended longer than regular storms, allowing hailstones to grow bigger. To form cricket-ball sized hailstones, there must be updrafts of around 160 km/h.

FIGURE 8 The force of a storm tore this tree from the ground.

FIGURE 9 In June 2016, another supercell storm hit Sydney. Waves up to eight metres high crashed into the shoreline at Collaroy Beach and caused extensive damage.

Both Adelaide and Sydney were pummelled by supercell storms on January 22. The worst hit areas were in the Adelaide Hills and Fleurieu Peninsula where 20 000 homes lost power and the SES responded to 61 calls for help. Thirty-five millimetres of rain was recorded in half an hour, resulting in flash flooding and hailstorms measuring two centimetres in diameter carpeting parts of the city. Wind gusts of up to 90 kilometres per hour were recorded at the airport.

Meanwhile, Sydney was warned to prepare for the worst, to secure vehicles and loose items, unplug electronic equipment and to stay indoors as the city braced itself for more storms, following from those experienced in previous days. The intense storm activity was the result of the large number of hot days. Flash flooding, damaging winds, hail and lightning were set to continue.

On 29 January, the tourist hot spots around the Gold Coast and Sunshine Coast were lashed by severe storm activity. Wind gusts of more than 100 kilometres per hour were recorded, with almost 9000 properties losing power.

5.5.6 How do I protect myself in a thunderstorm?

During storms, damage and injury are often caused by loose objects blown around by the wind, by lightning strikes, and by people being caught in flash floods. To protect yourself, take the following precautions:

- Before the storm approaches, make sure loose objects outside your home are secure.
- Stay inside during the storm.
- Unplug electrical equipment such as computers, televisions and gaming consoles.
- Avoid using the phone until the storm has passed.
- Use torches rather than candles as a source of light.
- Stay indoors, and stay away from windows.
- If caught in a storm, try to find shelter.
- If caught in the open, move away from objects that could fall, such as trees.
- Crouch down; don't huddle in a group.
- Never try to walk or drive through floodwater.
- Do not touch or approach fallen power lines.

FIGURE 10 The roof of a house sits in the middle of the road at The Gap in Brisbane's north-west. The roof is from a home 50 metres away.

 Resources

 Weblink The Gap storm

5.5 INQUIRY ACTIVITIES

1. Use the diagrams in this subtopic to make your own sketch of a supercell storm. Using words such as evaporation, condensation and precipitation, annotate your diagram to explain how storms develop.
   Classifying, organising, constructing

2. (a) Use the information in this subtopic to annotate a map of Australia to show the dates when thunderstorms were recorded around Australia and the damage they caused. Use the internet to find information to annotate **places** that are not mentioned in this subtopic.
   (b) Explain why so much thunderstorm activity occurs during January
   Classifying, organising, constructing

5.5 EXERCISES

Geographical skills key: GS1 Remembering and understanding **GS2** Describing and explaining **GS3** Comparing and contrasting **GS4** Classifying, organising, constructing **GS5** Examining, analysing, interpreting **GS6** Evaluating, predicting, proposing

5.5 Exercise 1: Check your understanding

1. **GS1** What is a thunderstorm?
2. **GS1**
   (a) List the *changes* to the *environment* and types of damage that might result from thunderstorm activity.
   (b) Next to each type of damage indicate:
      • whether the damage is caused predominantly by wind or water
      • whether the damage tends to occur to the natural or built *environment*.
3. **GS2** Study **FIGURE 7** and explain how hailstones are formed.
4. **GS2** Suggest reasons why people in earlier civilisations assumed weather events were the work of the gods.
5. **GS2** Explain why thunderstorms can cause so much damage to the natural and human *environments*.

5.5 Exercise 2: Apply your understanding

1. **GS2** Study **FIGURE 7**, which shows a supercell storm. Write a paragraph explaining why hailstones can vary so much in size.
2. **GS6** During which seasons of the year are thunderstorms more likely? Give reasons for your answer.
3. **GS5** Study **FIGURE 5**. During which hours of the day do most severe thunderstorms occur? Why?
4. **GS6** Select three points from the list of actions on how to protect yourself in a thunderstorm. Explain the rationale between the points you have chosen.
5. **GS6 FIGURES 8** and **9** show damage that resulted from thunderstorm activity. Refer to the Beaufort Scale (**FIGURE 7** in subtopic 5.2) and predict the wind speeds that might have been associated with this thunderstorm.

Try these questions in learnON for instant, corrective feedback. Go to www.jacplus.com.au.

5.6 SkillBuilder: Creating a simple column or bar graph

What are column or bar graphs?

Column graphs show information or data in columns. In a bar graph the bars are drawn horizontally and in column graphs they are drawn vertically. They can be hand drawn or constructed using computer spreadsheets.

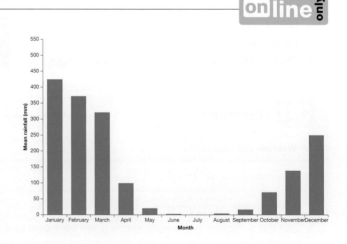

Select your learnON format to access:

• an overview of the skill and its application in Geography (Tell me)
• a video and a step-by-step process to explain the skill (Show me)
• an activity and interactivity for you to practise the skill (Let me do it)
• questions to consolidate your understanding of the skill.

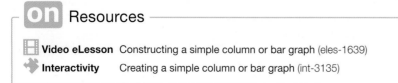

Resources

Video eLesson Constructing a simple column or bar graph (eles-1639)

Interactivity Creating a simple column or bar graph (int-3135)

5.7 Typhoons in Asia

5.7.1 Typhoon activity

In 2018 the western Pacific experienced an above average season of storm activity. Forty-five major systems were identified: 16 tropical depressions and 29 major storms. And of these, 13 intensified and developed into typhoons with a further 7 considered super typhoons. While some of these systems remained at sea, the collective damage bill was US$18.4 billion and 771 deaths.

TABLE 1 Typhoon activity from July to August 2018, the peak season for typhoons in Asia

Typhoon	Dates active	Area affected	Strongest winds (kph)	Fatalities	Damage ($US)
Maria (Gardo)	July 3–12	Mariana Islands, Ryukyu Islands, Taiwan, East China	195	2	$623 million
Sohn-Tinh (Henry)	July 16–24	Philippines, South China, Vietnam, Laos, Thailand, Myanmar	75	170	$256 million
Jongdari	July 23–August 4	Japan, East China	140	None	$1.48 billion
Shanshan	August 2–10	Marianna Islands, Japan	130	None	$132 000
Soulik	August 15–24	Carolina Islands, Mariana Islands, north-east China, Japan, Korean Peninsula, far-east Russia, Alaska	155	86	$84.5 million
Cimaron	August 16–24	Marshall Islands, Marianna Islands, Japan, Aleutian Islands	155	None	$30.6 million
Jebi (Maymay)	August 26–September 4	Marianna Islands, Taiwan, Japan, far-east Russia, Arctic	195	17	$3.29 billion

5.7.2 Typhoon Maria

Originally detected as a tropical disturbance over the Marshall Islands on 26 June 2018, the weather system was monitored for five days as it moved in a westerly direction. On 2 July, a Tropical Cyclone Formation Alert was issued by the Japanese Meteorological Agency. With sea surface temperatures fluctuating between 30 °C and 32 °C the system was upgraded to a tropical depression on 3 July when it was sitting off the coast of Guam. On 4 July it was upgraded to a tropical storm and named Maria. Within 24 hours Typhoon Maria reached typhoon status and was declared a super typhoon on 6 July when sustained wind gusts in excess of 260 kilometres per hour (kph) were recorded.

FIGURE 1 Super Typhoon Maria was referred to as Gardo, a local name, once it entered the Philippines' area of responsibility.

Source: NASA (2009)

Typhoon Maria made landfall at several places, losing intensity when crossing land and re-intensifying when moving back over water, before finally dissipating on 12 July.

On the island of Guam, the Anderson Airforce Base reported damage to aircraft, including air refuelling aircraft. Coastal communities suffered severe flooding and significant damage to infrastructure.

A severe weather warning was issued for Taiwan. As it was expected that Typhoon Maria would make landfall on Taiwan, schools, businesses and the airport were forced to close. However, communities on Taiwan were spared the full force of the typhoon when the eye of the storm passed to its north as it continued to track towards China. Taiwan was still lashed by heavy rain which led to flash flooding.

By the time Typhoon Maria crossed the Chinese coast 480 kilometres south of Shanghai it had been downgraded to a category 2 system, and it continued to weaken as it travelled inland from the coast.

Although only two fatalities were reported as directly caused by Typhoon Maria, it is thought that hundreds more deaths can be indirectly attributed to this event as a result of landslides and flooding that occurred in the wake of the storm.

FIGURE 2 Coastal areas in Japan are threatened by huge waves caused by Typhoon Maria.

FIGURE 3 Parts of Japan were inundated as a result of heavy rain from Typhoon Maria.

In groups of three or four, use the map in **FIGURE 1**, subtopic 5.4, to make a list of the countries most at risk from cyclones, hurricanes and typhoons. Thinking about the impact of heavy rainfall, storm surges and flooding, select the country you think might be most affected in terms of economic, social and environmental impacts. Use the internet to test your theory and outline these impacts. Present your findings to the rest of the class.

Evaluate your performance as a team member and assess how well you supported other members of your team. Write a short reflection on your performance. **Classifying, organising, constructing**

5.7 EXERCISES

Geographical skills key: GS1 Remembering and understanding **GS2** Describing and explaining **GS3** Comparing and contrasting **GS4** Classifying, organising, constructing **GS5** Examining, analysing, interpreting **GS6** Evaluating, predicting, proposing

5.7 Exercise 1: Check your understanding

1. **GS1** What is a typhoon?
2. **GS1** Which *places* were affected by Typhoon Maria?
3. **GS2** Does a typhoon need to cross the coastline to cause damage or loss of life? Explain.
4. **GS2** Typhoons weaken once they cross the coast. What major threats would be posed to environments away from the coast?
5. **GS1** Identify the four stages in the development of a typhoon.

5.7 Exercise 2: Apply your understanding

1. **GS6** Suggest at least two reasons why the typhoons discussed in this subtopic had different wind speeds and impacts.
2. **GS6** Explain why typhoons might weaken and re-form several times on their journey.
3. **GS6** Describe the *interconnection* between Typhoons Soudelor and Goni and the Mariana Islands, Taiwan, China and Japan.
4. **GS6** Explain how deaths that occurred away from the Chinese coast are thought to have been caused by Typhoon Maria.
5. **GS4** What do you think is the essential difference between a typhoon and a super typhoon?

Try these questions in learnON for instant, corrective feedback. Go to www.jacplus.com.au.

5.8 The impact of tornadoes on people and the environment

5.8.1 What is a tornado like?

Tornadoes (or twisters) are violent, wildly spinning columns of air that drop down from under a cumulonimbus cloud and make contact with the ground. They are different from cyclones in that they form over land rather than over water. Unlike cyclones, they are not dependent on a supply of warm water to keep them going. Use the **Tornadoes 101** weblink in the Resources tab to watch a video about how tornadoes form, the damage they cause and how to survive them.

Tornadoes are ranked using a scale called the Enhanced Fujita scale (see **TABLE 1**). Five categories of wind speed are estimated, based on the damage left behind. These are not wind speed measurements, because most wind speed measuring devices are destroyed during tornadoes, and because the tornadoes die out so quickly.

Widely used in the United States and Canada, the Enhanced Fujita scale is a new version of the original Fujita scale and was designed to expand the descriptions of damage caused at different wind speeds. It also includes descriptions of damage to both the natural and human environments. Within the human environment, the Enhanced Fujita scale includes reference to differences in the construction quality of buildings.

FIGURE 1 The anatomy of a tornado

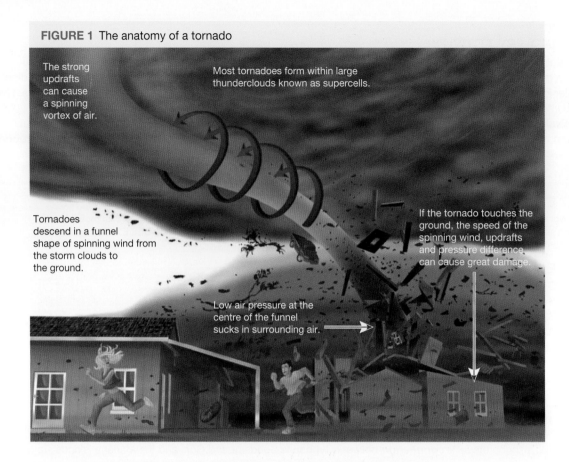

The strong updrafts can cause a spinning vortex of air.

Most tornadoes form within large thunderclouds known as supercells.

Tornadoes descend in a funnel shape of spinning wind from the storm clouds to the ground.

If the tornado touches the ground, the speed of the spinning wind, updrafts and pressure difference can cause great damage.

Low air pressure at the centre of the funnel sucks in surrounding air.

TABLE 1 The Enhanced Fujita scale, like the Beaufort scale, links tornado categories to the damage caused.

Scale wind speed (km/h) category	Typical damage	3-second wind gust speed (km/h)
E0 64–116 Gale	Some damage to chimneys; branches broken off trees; shallow-rooted trees pushed over; signboards damaged.	72–125
E1 117–180 Moderate	Peels surface off roofs; mobile homes pushed off foundations or overturned; moving autos blown off roads.	126–188
E2 181–252 Considerable	Roofs torn off frame houses; mobile homes demolished; train carriages overturned; large trees snapped or uprooted; light-object missiles generated; cars lifted off ground.	189–259
E3 253–331 Severe	Roofs and some walls torn off well-constructed houses; trains overturned; most trees in forest uprooted; heavy cars lifted off the ground and thrown.	260–336
E4 332–418 Devastating	Well-constructed houses levelled; structures with weak foundations blown away some distance; cars thrown and large missiles generated.	337–420
E5 419–512 Incredible	Strong frame houses levelled off foundations and swept away; automobile-sized missiles fly through the air in excess of 100 metres (109 yards); trees debarked.	421+

 Resources

🔗 **Weblink** Tornadoes 101

5.8.2 Where is Tornado Alley?

Tornadoes can occur anywhere, but most occur during spring and summer in a part of the United States known as **Tornado Alley** (see **FIGURE 2**). The worst tornado on record was the Tri-State tornado in March 1925. It destroyed towns across Missouri, Illinois and Indiana, killing 689 people.

The year 2011 ranks third in the history of the United States for the greatest number of strong to violent tornadoes. This number was surpassed only in 1974 and 1965.

FIGURE 2 A map of Tornado Alley, showing the areas of highest risk and high risk

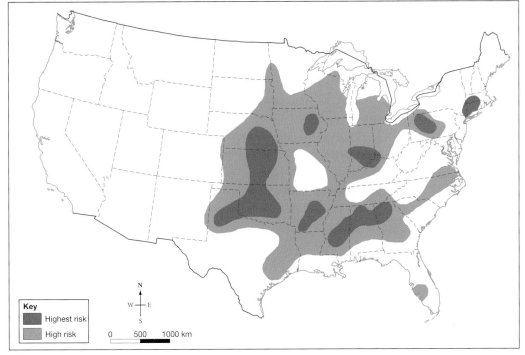

Key
Highest risk
High risk

0   500   1000 km

Source: MAPgraphics Pty Ltd, Brisbane

Data collected between 1979 and 2018 suggests that Tornado Alley may be experiencing a shift towards the east (see **FIGURE 3**). Scientists believe that this eastward shift is due to climate change. On average 1287 tornadoes are recorded across the United States each year and 70 deaths per year are directly attributed to tornadoes. One of the least active years on record was 2018, with only 991 tornadoes recorded.

Living in Tornado Alley

The people who live in Tornado Alley are well aware of the potential disaster that they face each year during spring and summer. Building codes have been strengthened, requiring all new buildings to have strong roofs and foundations that are tethered to the structure. Most neighbourhoods have early-warning sirens that sound when a tornado is imminent. Most homes have basements or underground **storm shelters** that provide protection for people during a tornado.

DISCUSS

Imagine you live in Tornado Alley in the US. Each spring and summer you face the dangers of a tornado. Prepare a ten-point checklist for your family that prepares them for this event. You need to account for the age and experiences of each family member, their strengths and weaknesses and the responsibilities they each should take in such an event.   **[Personal and Social Capability]**

FIGURE 3 Tornado frequency is increasing in eastern states such as Arkansas, Tennessee and Mississippi, but decreasing in traditional hot spots such as Texas.

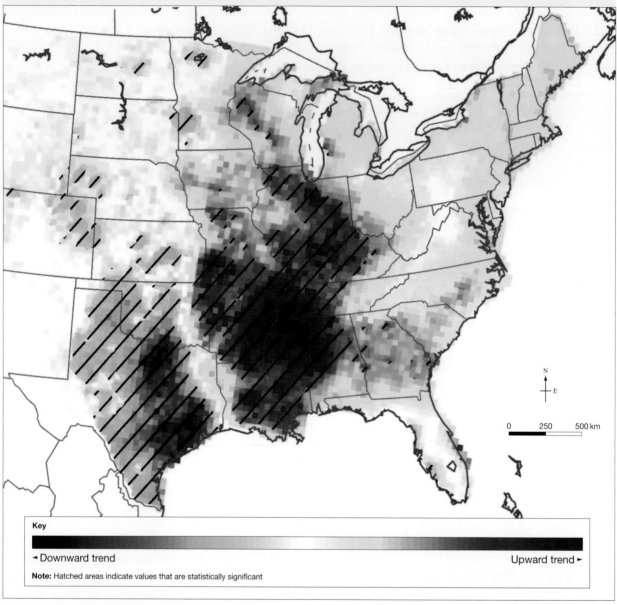

Key

◄ Downward trend   Upward trend ►

Note: Hatched areas indicate values that are statistically significant

0   250   500 km

Source: Gensini & Brooks (2018)

FIGURE 4 In April 2018 in Greensboro, North Carolina, trees were ripped from the ground, buildings flattened and electricity cut when a tornado carved a 54-kilometre path of destruction through the township. Wind speeds in excess of 217 kph were recorded.

5.8.3 CASE STUDY: New Orleans Tornado February 2017

Tornadoes can have a devastating effect on people and the environment. One such tornado descended on New Orleans, situated in the infamous Tornado Alley in the United States, on 7 February 2017. This tornado was the strongest recorded in New Orleans history.

A destructive supercell thunderstorm developed in New Orleans East on 7 February 2017. Supercell thunderstorms are very large storms that can last for many hours and are responsible for nearly all of the significant tornadoes that occur in the United States. They also produce most of the hailstones that are larger than golf-ball size (see **FIGURES 6** and **7** in subtopic 5.5). This tornado was one of several that touched down across the states of Mississippi and Louisiana.

The tornado formed when very high humidity, more closely associated with summer than winter levels, caused by warm water temperatures in the Gulf of Mexico collided with a trough of low pressure, which created atmospheric instability. As moisture was pushed ashore and into the path of the low-pressure trough it triggered a cycle of rapid up-draughts, which is the cooling and warming of the surrounding air. This continual cycle of air rising, condensing and falling created the perfect wind shear conditions for tornado development. Several tornado warnings were issued as forecasters monitored the development of this weather system and several others that threatened both Louisiana and Mississippi on 7 February. The path of the tornado is shown in **FIGURE 5**.

Covering a distance of 10.35 kilometres and with a width of 220 metres, the tornado touched down at 11.50 am northeast of Watson Louisiana. It left a trail of destruction before reaching its end point 10 kilometres east-north-east of Watson. Dozens of people were injured, cars were flipped, and buildings flattened or blown away. Around 12 000 homes were left without power. The damage bill was estimated at $2.7 million. The tornado had a damage rating of EF3 (see **TABLE 1**).

FIGURE 5 Path of the New Orleans Tornado 2017

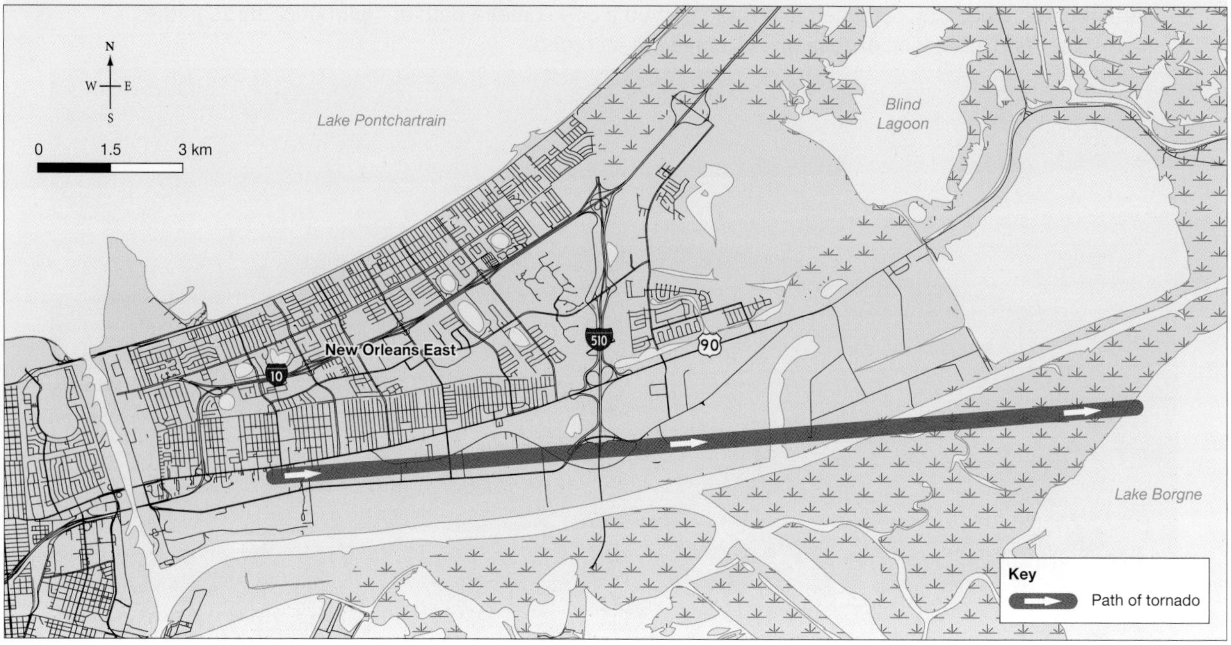

N
W—E
S

0   1.5   3 km

Lake Pontchartrain

Blind Lagoon

New Orleans East

510

90

10

Lake Borgne

Key

Path of tornado

Source: National Oceanic and Atmospheric Administration

FIGURE 6 Homes were destroyed and cars severely damaged by violent winds in the tornado that struck in February 2017.

END
SCHOOL
ZONE

on Resources

🌀 **Interactivity** Spiralling twister (int-3088)

🔗 **Weblink** Where tornadoes are likely to occur

Explore more with myWorldAtlas

Deepen your understanding of this topic with related case studies and questions.
- Investigate additional topics > Natural hazards > **Cyclones and tornadoes**

5.8 INQUIRY ACTIVITIES

1. Tornado Alley is well known, but it is not the only *place* in the world where tornadoes occur. Use the internet to find other locations where tornadoes occur regularly. Investigate one region and one tornado that has taken place. Using ICT, make a short movie clip of the event. Include information about the *scale* of the tornado. You must also cover the aftermath, detailing the impact on both the natural and built *environments*. **Classifying, organising, constructing**

2. Obtain a map of the area local to your school or home (e.g. from Google Earth). To understand the *scale* of the February 2017 storm, make your home or school the centre of the map area and measure 16.3 kilometres across. Describe the area affected and the damage that would have been caused if a similar tornado occurred in your area **Comparing and contrasting**

5.8 EXERCISES

Geographical skills key: GS1 Remembering and understanding **GS2** Describing and explaining **GS3** Comparing and contrasting **GS4** Classifying, organising, constructing **GS5** Examining, analysing, interpreting **GS6** Evaluating, predicting, proposing

5.8 Exercise 1: Check your understanding

1. **GS1** When are tornadoes most likely to occur? Give reasons for your answer.
2. **GS1** What is Tornado Alley and where is it?
3. **GS2** Refer to subtopic 5.4. Explain the difference between a cyclone and a tornado.
4. **GS5** Refer to **FIGURE 1**. What is the *interconnection* between large thunderclouds and tornadoes?
5. **GS2** Do you think we have tornadoes in Australia? Explain.
6. **GS1** What is a supercell thunderstorm?
7. **GS2** Describe the conditions that led to the formation of the February 2017 tornado in New Orleans.
8. **GS2** Describe the damage that results from an EF3 tornado.

5.8 Exercise 2: Apply your understanding

1. **GS5** Study the graph of tornado frequency in **FIGURE 7**.
   (a) During which hours of the day are tornadoes more likely?
   (b) Suggest a reason for this pattern.

▶

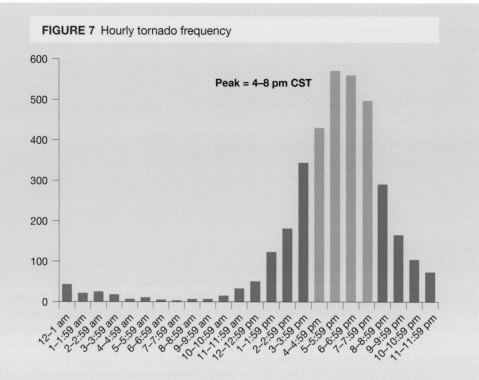

FIGURE 7 Hourly tornado frequency

Peak = 4–8 pm CST

2. **GS6** Imagine you are equipping a storm shelter. What ten items could your shelter not do without? Justify your choices.
3. **GS5** Refer to **FIGURE 2**. How does the *scale* of risk of tornadoes occurring vary within the United States?
4. **GS6** Using the information provided in the case study, estimate the area that was affected by this tornado.
5. **GS6** Describe the changes in tornado activity in Tornado Alley. In your answer refer to actual states and suggest a reason for this change.

Try these questions in learnON for instant, corrective feedback. Go to www.jacplus.com.au.

5.9 When water turns to ice and snow

5.9.1 What is a blizzard?

Periods of intense snowfall characterised by high winds and snow are known as snowstorms and can be just as deadly as any other storm. The most dangerous snowstorm of all is the blizzard.

The difference between a snowstorm and a blizzard is the strength of the wind. A snowstorm is officially recognised as a blizzard when wind speed is sustained above 56 kilometres per hour or has frequent gusts in excess of this speed for more than three hours. Visibility in a blizzard is also reduced to less than 400 metres. In the most extreme cases it may be difficult to see beyond a metre ahead. Often snow does not fall during a blizzard, but is blown into snowdrifts capable of burying people and objects.

What causes blizzards?

Variations in air pressure (see section 5.2.2) cause strong winds when warm air and cold air meet. It is these strong winds and cold conditions that cause a blizzard to develop.

FIGURE 1 How a blizzard forms

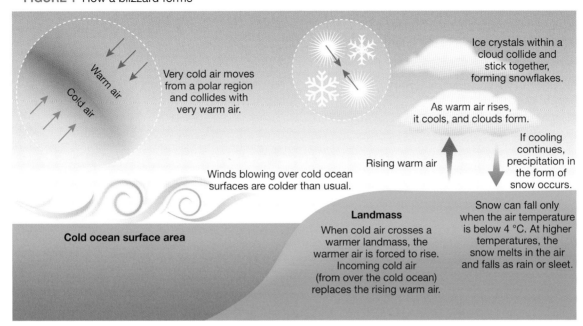

Very cold air moves from a polar region and collides with very warm air.

Warm air

Cold air

Ice crystals within a cloud collide and stick together, forming snowflakes.

As warm air rises, it cools, and clouds form.

Rising warm air

If cooling continues, precipitation in the form of snow occurs.

Winds blowing over cold ocean surfaces are colder than usual.

Cold ocean surface area

Landmass
When cold air crosses a warmer landmass, the warmer air is forced to rise. Incoming cold air (from over the cold ocean) replaces the rising warm air.

Snow can fall only when the air temperature is below 4 °C. At higher temperatures, the snow melts in the air and falls as rain or sleet.

Why are blizzards dangerous?

During snowstorms, snow can pile up, and it is impossible to know the depth of the snow, making it difficult to move about. There is the risk of falling through thin ice or into deep **crevasses**. Snow also tends to pile up on slopes. Where the snow load is greater than can be supported by the slope, there is a risk of **avalanches** (see **FIGURE 2**). An avalanche can be triggered by an earthquake or loud noises such as those produced by a gunshot or by animals. During blizzards a condition known as a **whiteout** can occur (see **FIGURE 3**). This means there is so much snow that visibility is severely affected and may be limited to just one metre. People and animals cannot tell the difference between the Earth and the sky, and quickly become disoriented, lose their way and risk freezing to death.

In the extreme cold associated with snowstorms and blizzards, people are at increased risk of **hypothermia**, **frostbite** and suffocation.

FIGURE 2 An avalanche

What was the world's deadliest blizzard?

The world's deadliest blizzard occurred in Iran in February 1972. A week of low temperatures and strong winds dumped more than three metres of snow and resulted in 4000 deaths. The weather conditions that led to this blizzard are shown in the satellite image in **FIGURE 4**.

FIGURE 3 Whiteouts reduce visibility, making it easy to become disoriented.

FIGURE 4 Weather conditions during the 1972 blizzard in Iran

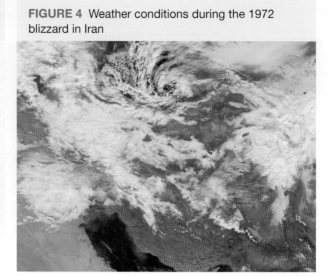

How can buildings be adapted to blizzards?

Researchers in Antarctica have to contend with snow build-up in some parts of the continent. The Halley VI facility (**FIGURE 5**) has been built on steel legs that can be raised. Skis have been attached to these legs, so that the entire station can be moved in order to eliminate the dangers associated with accumulating snow.

FIGURE 5 The Halley VI Research Station in Antarctica

5.9.2 CASE STUDY: How do blizzards affect the United States?

On 22–24 January 2016, a major blizzard dumped 91 centimetres of snow on parts of the mid-Atlantic and north-east United States. The snowstorm covered approximately 1.125 million square kilometres and affected more than 102 million people. It was officially rated as a category 4 snowstorm for the south-east and category 5 in the north-east. It is among the most powerful storms of all time. Snowfall records were set in a number of cities, including parts of New York, Pennsylvania and North Carolina.

TABLE 1 The Northeast Snowfall Impact Scale combines data on the area covered, the amount of snowfall and the population of the area.

Category	NESIS value	Description
1	1–2.499	Notable
2	2.5–3.99	Significant
3	4–5.99	Major
4	6–9.99	Crippling
5	10.0+	Extreme

FIGURE 6 A polar vortex was responsible for the blizzard, and brought record snow falls to the United States.

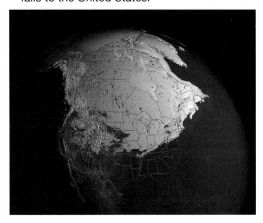

What caused such a severe blizzard in January 2016?

The January blizzard began as an atmospheric disturbance on 20 January that developed into a weak low-pressure system before intensifying and triggering a series of thunderstorms. Fed by a **polar vortex** (see **FIGURE 6**), air pressure continued to drop and the system developed into a major storm, bringing freezing rain, sleet and heavy snowfalls as it moved in a north-easterly direction across the United States (see **FIGURE 7**). The remnants of the storm were felt as far away as the United Kingdom before it finally dissipated over Finland on 29 January. At its peak, the storm system recorded a central air pressure of 983 millibars.

FIGURE 7 Eleven states in the United States were affected by the blizzard in January 2016.

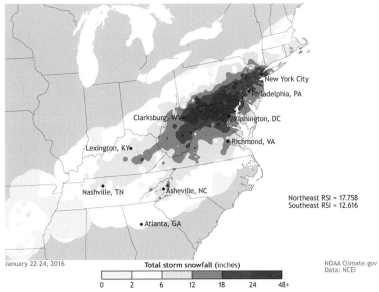

How did this blizzard affect people and places?

A state of emergency was declared in Washington DC and 11 other states as authorities prepared for record-breaking snowfalls. More than 100 million people were affected, with 33 million placed on blizzard watch. Around 500 000 people suffered power outages. The National Guard was placed on standby and crews were brought in from other parts of the country to assist power companies check lines and restore power. Around 1.5 million tonnes of road salt was distributed across the region and more than 7000 pieces of snow equipment were mobilised to help deal with the aftermath. People were urged to stay indoors and off the roads.

Records tumbled across the United States. The highest falls were recorded at Mount Mitchell in North Carolina with 170 centimetres of snow — a new all-time record (see **TABLE 2**). Daily snowfall records were also recorded in several places.

TABLE 2 All-time record snowfalls

Location	State	Amount
Mount Mitchell	North Carolina	170 cm
Allentown	Pennsylvania	81 cm
Philadelphian International Airport	Pennsylvania	77 cm
John F Kennedy International Airport	New York	77 cm
Baltimore–Washington International Airport	Maryland	74 cm
Newark	New Jersey	71 cm

The blizzard caused major disruptions for people travelling by air, with the travel plans of more than 100 000 travellers thrown into chaos (see **FIGURE 8**). More than 10 000 flights were cancelled in the United States, many of these due to the '**ripple effect**' caused by the cancellation of flights into and out of airports in the affected area. The number of cancelled flights also 'rippled' internationally, with around 200 flights in Canada, Mexico and the United Kingdom also cancelled.

The damage bill was estimated to be as high as US$3 billion. Fifty-five fatalities were also recorded; some of which have been directly attributed to heart attacks caused by shovelling snow. In Washington DC, police issued almost $1.1 million worth of parking fines and $65 000 in fines for cars abandoned on snow emergency routes. More than 700 vehicles were towed by authorities and impounded.

FIGURE 8 Flights across the affected region were grounded after record snow falls.

Why is shovelling snow dangerous?

In the United States, more deaths occur during blizzards than from hypothermia and motor vehicle accidents combined. Every winter around 100 people in the United States die from heart attacks caused by shovelling snow. Researchers have found that shovelling snow places more strain on the heart than a vigorous session on a treadmill. Both heart rate and blood pressure increase more dramatically when using arm muscles as opposed to leg muscles. Most people shovel snow in the early morning when the temperature is at its coldest, causing arteries to constrict which in turn decreases blood supply which in turn leads to cardiac arrest.

5.9 INQUIRY ACTIVITY

1. Refer to the work on natural hazards you have completed in other subtopics.
    (a) Explain which hazard you believe would be the most dangerous. Explain why.
    (b) Survey your class to find out which hazard is voted the most dangerous.
    (c) As a class, compile a list of reasons why class members voted for their chosen hazard.

Evaluating, predicting, proposing

5.9 EXERCISES

Geographical skills key: GS1 Remembering and understanding **GS2** Describing and explaining **GS3** Comparing and contrasting **GS4** Classifying, organising, constructing **GS5** Examining, analysing, interpreting **GS6** Evaluating, predicting, proposing

5.9 Exercise 1: Check your understanding

1. **GS1** How are snowstorms and blizzards different?
2. **GS2** Why are whiteouts so dangerous?
3. **GS2** Describe the visible *changes* to the *environment* that a blizzard brings.
4. **GS2** Where is the Halley VI Research Station? Explain why it has been built on skis.
5. **GS6** Suggest reasons for power outages during a blizzard.

5.9 Exercise 2: Apply your understanding

1. **GS6** Study **FIGURE 9**. Who do you think is most at risk from the extreme cold? Give reasons for your answer.

FIGURE 9 Who is at most risk from extreme cold?

2. **GS5** Describe the *interconnections* between the atmosphere and the land that cause blizzards.
3. **GS2** With the aid of a diagram, explain the operation of the water cycle in colder regions.
4. **GS5** Explain the *interconnection* between blizzards, shovelling snow and heart attacks.
5. **GS5** Refer to the case study on blizzards in the United States. The blizzard is said to have had a 'ripple effect' on people's travel plans. Explain what the term 'ripple effect' means in this context and how people were affected.

Try these questions in learnON for instant, corrective feedback. Go to www.jacplus.com.au.

5.10 Responding to extreme weather events

5.10.1 How do I prepare?

With today's modern technology we have access to a wealth of information that enables individuals and communities to prepare themselves for the wild winds over which they have no control. In many cases, the winds also bring vast amounts of rainfall and the land is often **inundated**. While we can in some ways prepare for such events, it is inevitable that both the natural and built environment will be affected.

People who live in disaster-prone areas should know the risks associated with the potential hazards they face and the time of the year when they are at greatest risk. In Queensland, for example, where tropical cyclones bring flooding rains, houses are often built on stilts.

The key to survival is to be prepared. Securing your home and having an emergency kit are two important things that can be done on a continual basis.

FIGURE 1 An unprepared home

(A) Overhanging branches
(B) Loose roof tiles
(C) Loose guttering
(D) Unsecured garden furniture
(E) Dangerous debris
(F) Unsecured children's toys

FIGURE 2 A well-prepared home

A No loose guttering
B Secure roof
C Trimmed branches
D Window shutters installed
E No unsecured items in garden

FIGURE 3 An emergency kit

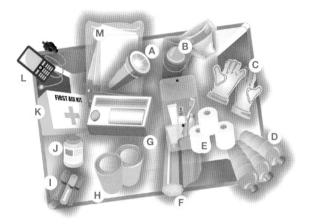

A Torch
B Baby formula and nappies
C Sturdy gloves
D Fresh water for three days
E Toiletries
F Waterproof bags
G Portable radio
H Three days' worth of non-perishable food and can opener
I Spare batteries for torch, radio and mobile phone
J Essential medication
K First aid kit
L Mobile phone and charger
M Important documents in sealed bags and cash

5.10.2 What happens in Chatham County?

Chatham County is located in the US state of Georgia. It has a population of approximately 270 000 and occupies an area of 1637.6 square kilometres.

On average, Chatham County expects up to 19 severe thunderstorms each year. As Georgia lies in Tornado Alley, people in Chatham can also expect an average of six tornadoes each season. In order to assist residents, a network of early warning sirens has been installed in places where people typically tend to gather. Ninety-five per cent of the county is covered. When the sirens give out their distinctive, extended wail, residents know to seek shelter. Most homes and public buildings in the county are equipped with storm cellars.

FIGURE 4 The location of sirens in Chatham County

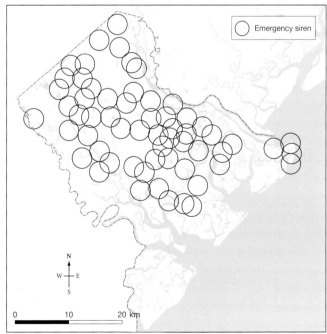

Source: Spatial Vision

FIGURE 5 The entrance to a storm cellar

DISCUSS

In Chatham County in Georgia, United States, 95 per cent of the county is covered by an early warning system to warn about severe thunderstorms. There are also homes and public buildings with storm cellars for people to seek shelter. Should it be the responsibility of individuals or governments to provide early warning and safe environments for people in such events? Establish your views and then ask the same question to members of your family. How do the views compare?   **[Ethical Capability]**

 Resources

🔧 **Google Earth** Chatham County

5.10 EXERCISES

Geographical skills key: GS1 Remembering and understanding **GS2** Describing and explaining **GS3** Comparing and contrasting **GS4** Classifying, organising, constructing **GS5** Examining, analysing, interpreting **GS6** Evaluating, predicting, proposing

5.10 Exercise 1: Check your understanding

1. **GS5** Study **FIGURES 1** and **2**. Explain to the people who live in the house shown in **FIGURE 1** what they need to do to prepare their home for the next cyclone season. Identify particular hazards and the potential risks they pose.
2. **GS2** Explain why the homes shown in **FIGURES 1** and **2** are elevated.
3. **GS6** What do you think poses the biggest threat — wind or water? Justify your choice.
4. **GS2** What have the people in Chatham County done to deal with hazards in their *environment*?
5. **GS5** What is the *interconnection* between roof slope and extreme weather events?

5.10 Exercise 2: Apply your understanding

1. **GS6** What do you think the people who live in the house in **FIGURE 2** are preparing for? Give reasons for your answer.
2. **GS6** The residents of the house shown in **FIGURE 2** live in a *place* that has been placed on cyclone alert and have asked you for advice. They have begun preparing an emergency kit like the one shown in **FIGURE 3**. Other than food, what items would you suggest they include?
3. **GS6** What items do you think you would need in a blizzard that you would not need in a thunderstorm, tornado or tropical cyclone? Explain.
4. **GS5** Which regions within Chatham County are not well covered by sirens? Use evidence from the map in **FIGURE 4** to suggest a reason for this.
5. **GS6** People without storm shelters, or who cannot evacuate before an extreme weather event, are advised to take shelter in a small room such as a bathroom. Suggest a reason for this.

Try these questions in learnON for instant, corrective feedback. Go to www.jacplus.com.au.

5.11 Thinking Big research project: Weather hazard documentary

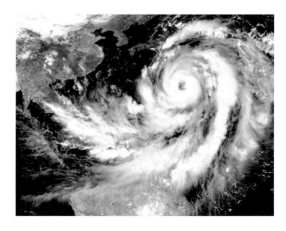

SCENARIO

The Bureau of Meteorology has issued a severe cyclone warning for northern Australia. Tropical cyclone Trevor is bearing down on the coast of far north Queensland and Tropical Cyclone Veronica is heading for the Western Australia's northern coast. You have been engaged by an independent media company to follow one of these cyclones and prepare a documentary that will air later this year.

Select your learnON format to access:

- the full project scenario
- details of the project task
- resources to guide your project work
- an assessment rubric.

Resources

projectsPLUS Thinking Big research project: Weather hazard documentary (pro-0236)

5.12 Review

5.12.1 Key knowledge summary

Use this dot point summary to review the content covered in this topic.

5.12.2 Reflection

Reflect on your learning using the activities and resources provided.

 Resources

 eWorkbook Reflection (doc-32137)

Crossword (doc-32138)

 Interactivity Natural hazards and extreme events crossword (int-7702)

KEY TERMS

avalanche rapid movement of snow down a slope, usually under the influence of gravity. It can also be triggered by animals, skiers or explosions.

barometer an instrument used to measure air pressure

crevasse a deep crack in ice

cumulonimbus clouds huge, thick clouds that produce electrical storms, heavy rain, strong winds and sometimes tornadoes. They often appear to have an anvil-shaped flat top and can stretch from near the ground to 16 kilometres above the ground.

cyclones intense low pressure systems producing sustained wind speeds in excess of 65 km/h. They develop over tropical waters where surface water temperature is at least 26 °C.

frostbite damage caused to the skin when it freezes, brought about by exposure to extreme cold. Extremities such as fingers and toes are most at risk, along with exposed parts of the face.

gale force wind wind with speeds of over 62 kilometres per hour

hailstone an irregularly shaped ball of frozen precipitation

hailstorm any thunderstorm that produces hailstones large enough to reach the ground

hypothermia a condition in which a person's core body temperature falls below 35 °C and the body is unable to maintain key systems. There is a risk of death without treatment.

intensify to become stronger

inundate to cover with water, especially floodwater

isobars lines on a map that join places with the same air pressure

meteorologist a person who studies and predicts weather

polar vortex a large pocket of very cold air rotating in the same direction as the Earth's orbit

ripple effect the flow-on effect of a particular action

storm shelter underground shelter where people can take refuge from a tornado

storm surge a sudden increase in sea level as a result of storm activity and strong winds. Low-lying land may be flooded.

Tornado Alley a region of the central United States across which tornadoes are most likely to form. The core states are Texas, Oklahoma, Kansas, Nebraska, eastern South Dakota, and the Colorado Eastern Plains.

torrential rain heavy rain often associated with storms, which can result in flash flooding

troposphere the layer of the atmosphere closest to the Earth. It extends about 17 kilometres above the Earth's surface, but is thicker at the tropics and thinner at the poles, and is where weather occurs.

typhoon the name given to cyclones in the Asian region

whiteout a weather condition where visibility and contrast is reduced by snow. Individuals become disoriented as they cannot distinguish the ground from the sky.

UNIT 2
PLACE AND LIVEABILITY

Have you ever stopped to consider why you live where you do? What prompted your family to live there? There are so many different types of places where you *could* live: rural or urban, coastal or inland, small or large, bustling or quiet. Different people find different places suitable (or more 'liveable') for them than other places. Some people have no choice. The question is: how can we make places more liveable?

GEOGRAPHICAL INQUIRY: WHAT IS MY PLACE LIKE?

Task

Create a blog that presents demographic characteristics of a local place. The ABS website (www.abs.gov.au) provides a pathway for you to find out the demographic characteristics of your chosen postcode area.

Select your learnON format to access:

- an overview of the project task
- details of the inquiry process
- resources to guide your inquiry
- an assessment rubric.

 Resources

ProjectsPLUS Geographical inquiry: What is my place like? (pro-0144)

6 Choosing a place to live

6.1 Overview

Australia has so much unused space. Why doesn't everyone spread out more?

6.1.1 Introduction

Why does your family live in the place, state, city, street or house that it does? Why do many Australians live in big cities near the coast? Have you thought about the reasons why your parents selected the place or environment in which you now live? People living in Australia have been making choices about where to live for many thousands of years. Has the region of Australia around Port Jackson in modern Sydney always been the most heavily populated part of Australia? Do people choose to live in places they feel are the most liveable? Let's try to work out why Australians choose to live in the places they do.

on Resources

- ☑ **eWorkbook**   Customisable worksheets for this topic
- 🎞 **Video eLesson**   Choosing a place to live (eles-1619)

To access a pre-test and starter questions and receive immediate, **corrective feedback** and **sample responses** to every question select your learnON format at www.jacplus.com.au.

6.2 A sense of place

6.2.1 What creates a sense of place?

Places are central to the study of geography. This is because geographers are interested in where things are found on Earth and why they are there. But what exactly is a place?

To understand what a place is, think about **location** and **region**. Each place has a unique identity that makes it different from other places. A combination of characteristics is specific to that place, making it individual. A sense of place comes from being aware of what makes that location significant and seeing its special qualities.

The characteristics of a place can come from:

1. natural features
2. human features — that is, features built by people
3. a combination of the two.

Eventually, one or more of these features becomes a symbol of that place in people's minds.

FIGURE 1 The Taj Mahal in Agra, India

FIGURE 2 The Grand Canyon, Utah, United States

FIGURE 3 Rio de Janeiro, taken from a helicopter, showing the Corcovado in the foreground with the statue of Christ on it and Sugarloaf Mountain, or Pao de Acucar, in the background, to the right

FIGURE 4 Disney World, Orlando, Florida, United States

FIGURE 5 Table Mountain, Cape Town, South Africa

FIGURE 6 The Golden Gate Bridge, San Francisco Bay, United States

FIGURE 7 The Great Wall of China

Explore more with my**World**Atlas

Deepen your understanding of this topic with related case studies and questions.
- Investigating Australian Curriculum topics > Year 7: Place and liveability > **New York City**

 on Resources

Google Earth Taj Mahal

Grand Canyon

Disney World

Table Mountain

San Francisco Bay

6.2 INQUIRY ACTIVITY

1. Conduct a survey of your class to find out each person's top five favourite *places* in Australia.
   (a) Collate the results in a table like the one below.

Place	Student A	Student B	Student C
Great Barrier Reef	✓		✓
Uluru		✓	✓
My grandparents' farm near Ballina, NSW		✓	

This table could also be set up electronically, using a spreadsheet program.

   (b) Graph the results to show the ranking of the places by percentage of the class; for example, 45 per cent of the class named Uluru in their top five places in Australia.
   (c) As a class, discuss the patterns shown by the graph. Suggest reasons to explain why people like or dislike certain places.

Classifying, organising, constructing

6.2 EXERCISES

Geographical skills key: GS1 Remembering and understanding **GS2** Describing and explaining **GS3** Comparing and contrasting **GS4** Classifying, organising, constructing **GS5** Examining, analysing, interpreting **GS6** Evaluating, predicting, proposing

6.2 Exercise 1: Check your understanding

1. **GS1** Define the following terms in your own words.
   (a) Place
   (b) Location
   (c) Region
2. **GS2** Study **FIGURES 1, 2, 3, 4, 5, 6** and **7**. Describe five characteristics in the *environment* of each feature that create its individual sense of *place*.
3. **GS3** Of all these characteristics, which one do you believe to be the most important in creating an identity for that *place* in the minds of people?
4. **GS5** Study **FIGURES 1, 2, 3, 4, 5, 6** and **7**. For each figure, identify whether the characteristics of the place come from natural features, human features, or both.
5. **GS6** Suggest reasons why these *places* have become famous around the world.

6.2 Exercise 2: Apply your understanding

1. **GS6** Do you think that people's favourite *places* would vary with the age of the individual? Explain your answer.
2. **GS6** No matter where we live, we all live in the one *place*: Planet Earth. From what you have learned so far, define what a *place* is in your own words. What do you think would be the characteristics of a *place* that would appeal to anyone, wherever they come from? (*Hint:* What feelings do you have when you are in a *place* that you like?)
3. **GS4** Name and describe a place you have visited and enjoyed that is predominantly made up of natural characteristics.
4. **GS4** Name and describe a place you have visited and enjoyed that is predominantly made up of human characteristics.
5. **GS4** Name and describe a place you have visited and enjoyed that is made up of both natural and human characteristics.

Try these questions in learnON for instant, corrective feedback. Go to www.jacplus.com.au.

6.3 Why people live in certain places

6.3.1 Push and pull

People choose to live in specific places for a wide range of reasons. These reasons can be broadly divided into **pull factors** and **push factors**. The combination of reasons varies from person to person, and what is an advantage for one person may be seen as a negative by someone else.

It is also true, though, that the reasons people choose to live in a place often change over time. Sometimes, these reasons might even be connected to the very existence of the place — or its changing nature.

There are four factors that influence the liveability of places or why people decide to live there:

1. available resources (money)
2. employment
3. relationships with other people (for example, wanting to be near family or moving for a partner's job)
4. lifestyle.

Many of these factors change throughout a person's life. For example, where a 20-year-old single person wants to live is often quite different from where someone in their forties, or someone with a partner and two teenage children, may want to live.

In other situations, the reason for living in a place may disappear. The town of Rawson, near Mt Erica in Victoria, was built for the people building the Thomson Dam in the 1970s and early 1980s (see **FIGURE 1**). After the project was finished in 1983, nearly every family left the town because there was no longer any work there. Its **community** identity

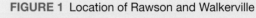

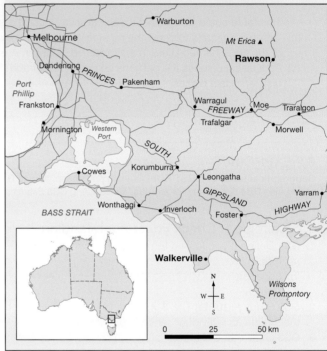

FIGURE 1 Location of Rawson and Walkerville

Source: Spatial Vision

had to change. The few people left in Rawson now provide services for people using the area for recreations such as bushwalking and skiing.

Walkerville is a small coastal settlement on the coast of Victoria, just east of Inverloch and Venus Bay near Wilsons Promontory (see **FIGURE 1**). Walkerville is a good example of the way people's reasons for living in a place can change over time.

Walkerville, settled in 1885, provided a place for workers during the early twentieth century who produced quicklime from the limestone cliffs. Cement was in great demand for building in Melbourne at this time, and lime could be transported there easily by ship. The town itself disappeared when the limestone cliffs were all mined out.

DISCUSS

Discuss the difficulties that would have been faced by the lime-burners who lived in the original settlement of Walkerville, given its *location*.

The modern settlement of Walkerville is now a small, isolated holiday location, popular with fishermen, and located next to the Cape Liptrap Coastal Park. Much of the original settlement of Walkerville no longer exists, but the ruins of the lime-burning kilns and the old cemetery still remain (see **FIGURE 2**).

FIGURE 2 (a) The Walkerville lime-burning kilns in 1972. (b) The kilns are now a tourist attraction. (c) There are no shops for 20 kilometres in either direction, except for the caravan park kiosk on the foreshore.

(a)   (b)   (c)

Many of the towns in the north-eastern United States were established as manufacturing towns. At first they were located near major ports or iron ore and coal deposits, and some closed down when these resources ran out. In more recent times, factories such as the one shown in **FIGURE 3**, which is near Baltimore, have closed down because the owners could no longer compete with the goods produced at a lower cost in China and other South-East Asian countries. With no other jobs available, people left the area, and it fell into a state of **urban decay**. In 2018, Baltimore is once again thriving, especially with many STEM jobs.

FIGURE 3 A disused factory near Baltimore

 **Resources**

     Interactivity   Push/pull factors (int-3089)

     Google Earth   Baltimore, USA

                          Walkerville

6.3 INQUIRY ACTIVITIES

1. Survey the members of your class and find out the reasons why their families chose to live in the *place* or *location* where they do. Classify the responses using the four categories named in this subtopic in a table like the one below.

Student	Resources	Employment	Relationships	Lifestyle
Gina	Near major shops	Near my dad's work		
Miguel		Near my mum's work	Close to my family who came to Australia earlier Close to my father's best friend	
Daniel				Near the sea, as we all sail or surf

(a) Present the answers using a column graph, correctly and fully labelled.

(b) As a class, discuss the pattern of reasons shown by the graph, and the possible explanations for this. For example, how important to people are social connections?

Examining, analysing, interpreting

2. Look up on Google Earth the *location* of the current settlement of Walkerville. Calculate the distance between Walkerville and the settlements around it. Study the land use and features of the environment around the settlement. Identify and list the advantages and disadvantages of Walkerville as a holiday *location*, using evidence from your Google Earth study. **Examining, analysing, interpreting**

6.3 EXERCISES

Geographical skills key: GS1 Remembering and understanding **GS2** Describing and explaining **GS3** Comparing and contrasting **GS4** Classifying, organising, constructing **GS5** Examining, analysing, interpreting **GS6** Evaluating, predicting, proposing

6.3 Exercise 1: Check your understanding

1. **GS2** Explain the difference between *push* and *pull* factors.
2. **GS1** What are four factors that influence the liveability of places and why people decide to live where they do?
3. **GS5** Identify and justify the push and pull factors that exist for people thinking about whether they should move to Walkerville today.
4. **GS2** Define the term 'urban decay' in your own words.
5. **GS5** Study **FIGURE 3**. Identify some of the specific signs that indicate an area is in urban decay.

6.3 Exercise 2: Apply your understanding

1. **GS6** Suggest reasons why some people continue to live in decaying urban *environments*, and why others might choose to move.
2. **GS3** Study **FIGURE 2 (c)** and think about the *changes* over *time* that have occurred in Walkerville as a *place*. Was the decline of the original township of Walkerville due to push or pull factors? How did these influence people's choice of where they would live? Justify your answer.
3. **GS6** A developer has proposed to the local shire council and the state government that the farmland around Walkerville should be rezoned to allow the building of a large holiday resort. In your opinion, would this be a good or bad policy for the future of the residents of Walkerville? Give reasons for your answer.
4. **GS5** Study **FIGURE 1** and answer the following two questions.
   (a) If you are in Leongatha, in what direction is Walkerville?
   (b) Use the scale to measure the straight line distance between Walkerville and Cowes.

Try these questions in learnON for instant, corrective feedback. Go to www.jacplus.com.au.

6.4 Where do you live and why?

6.4.1 How did I get here?

When we first learned to write our address, we often included our house number, street, town, city, state, country, continent, hemisphere, planet and universe. You could also identify your location with GPS coordinates, a grid reference or by use of latitude and longitude. Knowing the place where you are is important, but so is how you got there.

FIGURE 1 Australians born overseas, 2016

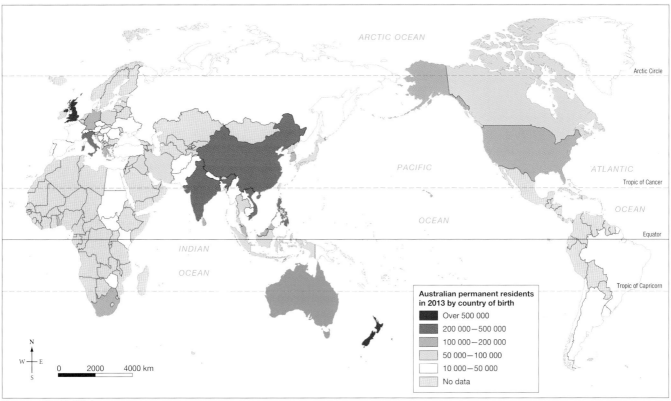

Source: Spatial Vision

A   *Cindy*: When I lived in Beijing I was called Jing-Wei. I came to Sydney in 2008 to study economics at university. I became an Australian citizen in 2013. I now have three Australian children.

B   *Andrew*: I came to Perth with my wife and three children in 1993 just before Nelson Mandela was elected president of South Africa. We were concerned for our safety in Johannesburg and were keen to start a new life in a country with a similar climate and language. Now two of my brothers also live in Australia.

C   *Lucy*: My brother moved from Palmerston North in New Zealand to Melbourne for work in 2006 and I followed him the next year. I like living in a larger city. There is more going on and I get paid a lot more. One day I might return to New Zealand.

D   *Deepak*: My family moved from Delhi in 1988 when my father was offered a job in a computer company in Adelaide. There were not many Indian kids in my school but I studied hard and went to university. I now have three children and live in Newcastle.

Did you and your family arrive by boat, plane or car, or were they born here? What decisions were made by your parents or grandparents which resulted in your family living in your place, house, state, country or hemisphere? In 2016, 28.5 per cent (nearly 7 million) of Australia's population was born overseas, and it is estimated that most will move homes several times times during their lifetime.

What is your story?

Activity 6.4, step 1 allows you to investigate why you live in your place. It is a task of discovery, and will take you some time to complete. Your aim is to discover your family's migration story and why you live where you do. Does your family have a recent migration story or did your family migrate with the First Fleet? Do you have Aboriginal or Torres Strait Islander heritage? Did your grandparent build the house you live in or did your parents or carers build your house? How has the place your family lives in changed over time and space?

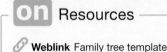 **Resources**

🔗 **Weblink** Family tree template

6.4 INQUIRY ACTIVITY
Your Family

This task allows you to discover your family's migration story and why you live where you do.

Step 1

What can you find out for yourself?

Your place

- What is your address? Write out your full address, including your hemisphere, latitude and longitude.
- Use Google Maps or Whereis to locate and identify your house in your street. Download an aerial view and a street view of your house.
- Annotate your aerial photo or map to identify who lives in your house, including pets, and which parts of the house they use. You could illustrate the people who live in your house in a cartoon — like the stickers, or decals, of families that people put on their cars.
- Ensure that your map has a compass, approximate scale and appropriate title.

Step 2

Your neighbourhood

- Using Google Maps or Whereis, download an aerial view of your street or at least the eight closest houses or dwellings.
- Using family decals, annotate each house to show who lives in it.

Step 3

How long have you lived at this address?

If you have previously lived somewhere else, list and map your past addresses. How many times have you moved? Share with your class the information that you have collected so far.

Step 4

How did you get here?

To investigate the rest of your story, you will need to speak to your parents and possibly your grandparents.
As you collect information about where your parents and grandparents were born, create a family tree of *places*. Try to find out why and when your relatives came to Australia. **FIGURE 2** illustrates how this may look.

- Where were your parents/carers born?
- How did they travel from where they were born to the *place* you now live?
- Why did your parents move to where you now live? Would they prefer to live in another place that is more liveable?
- Why did your grandparents and great-grandparents move from their *place* of birth?

Use the **Family tree template** weblink in the Resources tab to create your family tree.

FIGURE 2 An example of what your family tree may look like

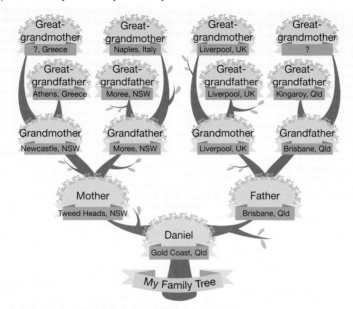

Use your family tree, along with the transport your family used, to create a map that shows this ***interconnection***.

Things to think about before starting your map:

- What ***scale*** and size of map will you need?
- Would you be better off having two maps? In the example shown in **FIGURE 2**, Daniel's parents and grandparents mostly came from New South Wales and Queensland, but most of his great-grandparents came from Europe. To map this information, he should use a world map plus a larger-***scale*** break-out or inset map of New South Wales and Queensland.
- How will you show the type of transport? Coloured arrows might work well.
- What is an appropriate title for your map?
- Would you like to illustrate your map with images of your relatives, their houses, flags of the countries they came from or images of the transport that they used? If you wish to add images, you will need to have a larger ***scale*** base map than if you just used symbols.
- You could annotate the map with the reasons your relatives moved.

FIGURE 3 Immigrants arriving in Australia by plane, 1967

6.4 EXERCISES

Geographical skills key: GS1 Remembering and understanding **GS2** Describing and explaining **GS3** Comparing and contrasting **GS4** Classifying, organising, constructing **GS5** Examining, analysing, interpreting **GS6** Evaluating, predicting, proposing

6.4 Exercise 1: Check your understanding

1. **GS5** Examine **FIGURE 1**. In 2016, in which countries were the largest number of Australian permanent residents born?
2. **GS5** Examine **FIGURE 1**. In 2016, in which countries were the smallest number of Australian permanent residents born?
3. **GS1** Complete the following sentence.
   In 2016 over _____ per cent of Australia's population was born overseas.
4. **GS1** It is estimated that most people will move homes _____ times during their lifetime. ▶

6.5 Australians living in remote places

6.5.1 Settling inland Australia

For over 100 years, a small percentage of Australians have been moving away from large cities and coastal regions to live in more **remote** locations. They are often searching for new farmland or the mineral resources of the inland. Why do some people choose to live in places where their nearest neighbour is 50 kilometres away and it takes six hours to get to the closest supermarket? Why do they find remote places more liveable?

The potential to relocate people inland has never been faster or easier. The interconnection provided by modern transport and the high-speed communication provided by phone and internet should mean that technology has reduced remoteness.

The general shift of Australia's population for the last 100 years has been towards the major cities and away from the country. In 2016 the average age of farmers in Australia was 56 years and getting older (this is up 17 years from 2010). Most children of farmers leave the country and seek education and work opportunities in large cities. **FIGURE 1** shows how quickly the inland of Australia was occupied after 1825.

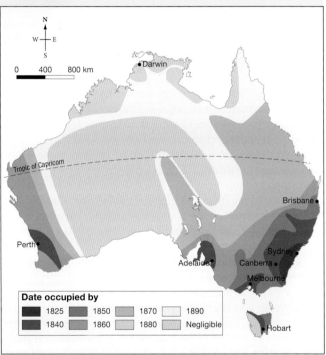

FIGURE 1 Stages in European land occupation in Australia

Date occupied by

1825	1850	1870	1890
1840	1860	1880	Negligible

Source: © Spatial Vision

Over the past 100 years, there have been many attempts by governments and private industry to encourage people to occupy the more remote places of Australia. Soldier settlement programs and mining developments are two such schemes.

Soldier settlement schemes

After both World War I and World War II, the state and federal governments of Australia began a program of providing land to returned soldiers. This was to give these soldiers work, but it was also seen as a way of attracting people to otherwise sparsely inhabited places.

After World War I, more than 25 000 soldiers were resettled in places such as Merbein and Mortlake in Victoria, Griffith and Dorrigo in New South Wales, Murray Bridge and Kangaroo Island in South Australia and the Atherton Tableland in Queensland. The settlers were expected to stay on their land for at least five years and to improve the quality of the land they were farming. Many of these settlements were not successful because the soldiers were not always suited to farming, the farms were often too small, and farmers did not have enough money to invest in stock or equipment.

After World War II, a similar scheme was much more successful, because farms were bigger and roads, housing and fences were supplied.

Remote mining communities

Karratha Broken Hill and Tom Price are examples of current mining towns that are just as remote as were the goldrush towns of Bathurst and Ballarat in the 1850s and 1860s.

Today it takes just over seven hours to fly from Melbourne to Tom Price, yet it can be difficult to attract workers to mines in this region. Wages are high; workers in the mining and construction industry in these locations can earn between $90–120 000 per year. There are now fewer jobs because the mining boom has passed, but skilled workers are attracted to these remote places. Some workers **fly in and fly out (FIFO)** for their shifts. In 2017 it was estimated that between 75 000 and 90 000 Australians fly in for a shift that may last several weeks, eventually flying home for their days off.

FIGURE 2 Location of soldier settlement areas, 1917

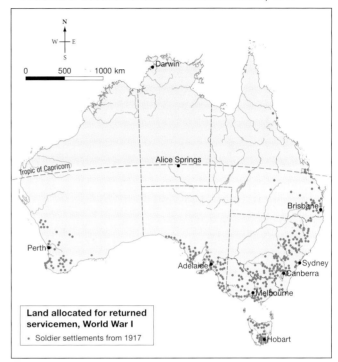

Source: Spatial Vision

FIGURE 3 Location of remote mining regions

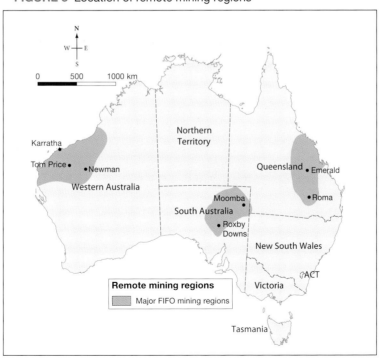

Source: Spatial Vision

FIGURE 4 Mount Tom Price mine and Tom Price township: note the FIFO workers' huts in the left foreground

6.5.2 Rural settlement

Some people live in rural areas because they are involved in primary industries. Others provide services.

Griffith is a large town (population 17 000) in the Murrumbidgee Irrigation Area in New South Wales. The climate in this area is semi-**arid** (warm, with unreliable rainfall). The land became productive farmland after **irrigation** was provided in 1912. Reliable water and available farmland attracted many people to this area.

FIGURE 5 Farms in the Griffith area support businesses in the town.

There are two main types of farm in this area.
- Type A farms are usually about 220 hectares in size (a hectare being 10 000 square metres). Each year they grow a combination of rice, corn, wheat, vegetables and pasture, and graze beef cattle. Irrigation water is usually used.
- Type B farms are **horticulture** farms, and are usually about 20 hectares in size. They grow a combination of permanent crops that may include grapes, peaches, plums, and citrus fruit such as oranges. Many of these plants last for many years, and irrigation is always needed.

FIGURE 6 Topographic map extract of Griffith

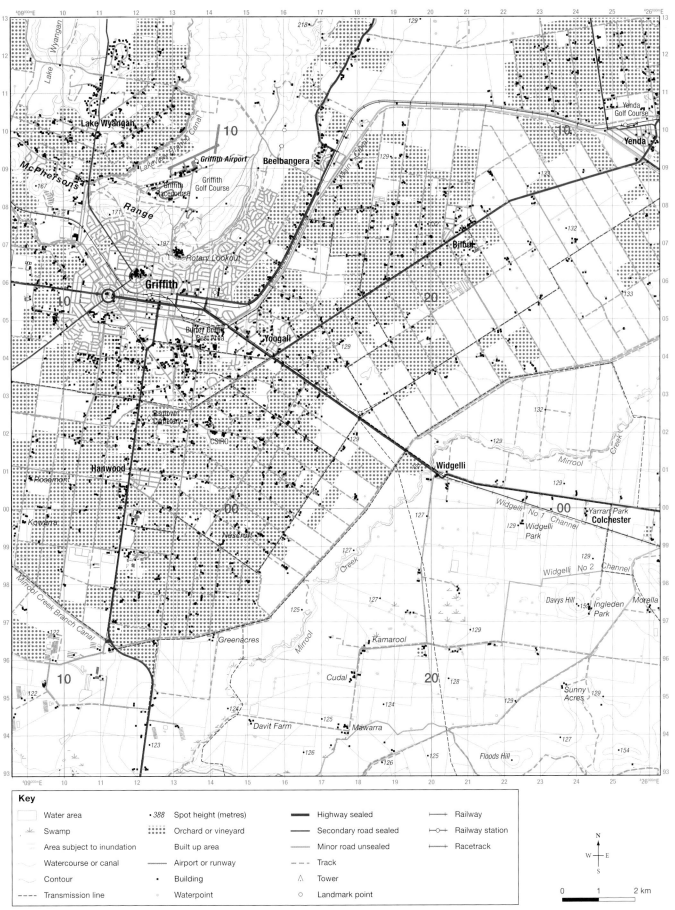

Key

▢ Water area	• 388 Spot height (metres)	▬▬ Highway sealed	⊢┼⊣ Railway	
⊥ Swamp	⦂⦂⦂ Orchard or vineyard	▬▬ Secondary road sealed	⊢○⊣ Railway station	
Area subject to inundation	Built up area	▬▬ Minor road unsealed	⊢⊣ Racetrack	
Watercourse or canal	▬▬ Airport or runway	▬ ▬ Track		
Contour	• Building	△ Tower		
▬ ▬ Transmission line	• Waterpoint	○ Landmark point		

N
W—E
S

0   1   2 km

Source: Spatial Vision

6.5.3 Are rural communities sustainable?

Rural communities are an important part of Australia's social identity, but they are facing significant change and challenges in maintaining their population. Many are experiencing a decline because young people are leaving in search of education and employment. Some rural communities are able to alter this trend, and are surviving against the odds. Others have not fared so well.

Coober Pedy is a vibrant multicultural town in the far north of South Australia, 850 kilometres north of Adelaide and 700 kilometres south of Alice Springs. The town is located in one of the most arid environments of Australia.

FIGURE 7 Coober Pedy location map

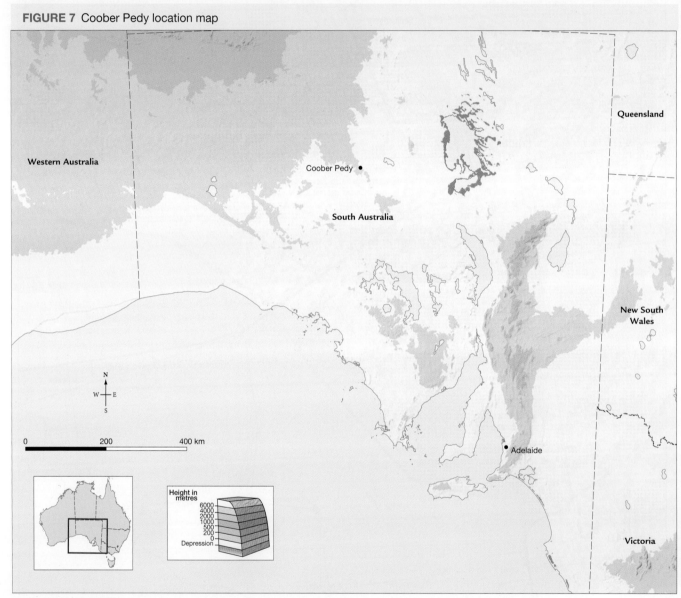

Source: Spatial Vision

For thousands of years, Aboriginal peoples walked the stony desert of the Coober Pedy area as part of their rotational occupation of land. The traditional custodians of the land are the Antakirinja people. The town's name may come from the name Kupa Piti, meaning 'white man's hole'. Opal was discovered in February 1915 and, after several cycles of boom and bust, the town expanded rapidly during the 1960s. Opal developed into a multi-million dollar industry, and the town is sometimes called the 'Opal Capital of the World'.

Opal continues to be important to Coober Pedy's identity and economy, but the town now draws its income from mining services, tourism and public services. Coober Pedy has a large Aboriginal community, and the town's population has now declined to an estimated 1762 people in the 2016 Census.

6.5.4 What does the future hold?

Coober Pedy is widely known for its underground housing (see **FIGURE 10**), an effective and environmentally friendly response to the town's searing summer heat and chilly desert evenings. Recent exploration has revealed significant deposits of iron ore, copper, gold and coal in the area, along with platinum, palladium and rare earths. Yet in 2014, the Cairn Hill iron ore/copper/gold mine was closed due to low iron ore prices.

The location of the town makes it an ideal centre for mining services, and a base for the delivery of state and federal government services in the region. This presents an opportunity for the town to reverse its steady population decline and again see growth in its economy and population.

Coober Pedy has good hospital and medical services, primary and secondary schooling, a TAFE campus, childcare services and police. However, these services are under some pressure, and there is a continuing problem with the recruitment and retention of medical professionals. This rural environment is extremely remote, so many of the **pastoral** properties in the region have been linked to telecommunication services since 1987. The Stuart Highway provides the main transport and service route for the town.

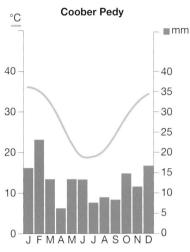

FIGURE 8 Climate graph for Coober Pedy, South Australia

FIGURE 9 Along with other South Australian fields, Coober Pedy produces most of the world's opal. Mullock heaps create Coober Pedy's distinctive landscape.

FIGURE 10 Much of Coober Pedy lives underground to take advantage of the cooler underground climate.

DISCUSS

Discuss strategies that could be implemented to entice more people to live in Coober Pedy and reverse its population decline. Develop a list of possible strategies that could be implemented.

[Critical and Creative Thinking Capability]

6.5.5 A question of survival

Many rural communities are facing global pressures, such as more overseas competition and fluctuations of the Australian dollar, which can affect the prices of commodities (minerals, wool, beef etc.). Climate change and resultant droughts and floods also have an impact on these rural communities. The rural communities that are not experiencing the trend of people moving to urban areas (**FIGURE 11**) all have one thing in common: they have discovered another source of income. They may have shifted their focus to growing olives or grapes, or perhaps made use of a natural environmental resource such as a nearby national park.

In some cases, a rural community is unable to reinvent itself or tackle the problem effectively. The loss of an industry such as mining may have terrible effects on employment, leaving the resident population with lower incomes and few job prospects.

FIGURE 11 Australia's population increase and decrease 2011–2016

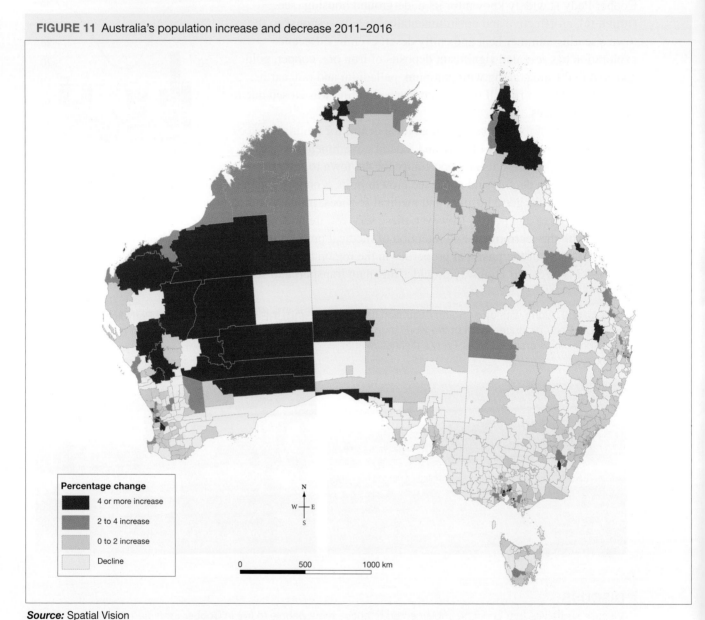

Percentage change

- 4 or more increase
- 2 to 4 increase
- 0 to 2 increase
- Decline

0   500   1000 km

Source: Spatial Vision

┌─ Explore more with my**World**Atlas ─────────────────────────────

Deepen your understanding of this topic with related case studies and questions.
- Investigate additional topics > Population > **Population of Australia**

Resources

Digital document Topographic map extract — Griffith (doc-17952)

Interactivity Remote living (int-3090)

Weblink Soldier settlement

Google Earth Tom Price township

Griffith

6.5 INQUIRY ACTIVITIES

1. Research a local soldier settlement scheme. When was it established? How successful was it? How did this scheme help to populate a remote *place*? Map its geographic features by using Google Maps. Use the **Soldier settlement** weblinks in the Resources tab to help with your research. **Describing and explaining**

2. How might people be encouraged to move from the coastal fringe to the more remote *places* of Australia? What could make you or your family move or relocate? Produce a short film, a snappy slide show or an advertising campaign that highlights the pull factors that might make people *change* the *place* where they live. **Evaluating, predicting, proposing**

3. Use an internet mapping tool to view a satellite image of Coober Pedy. What clues can you see that tell you about the climate of Coober Pedy? Can you see the mullock heaps, as shown in **FIGURE 9**? **Examining, analysing, interpreting**

6.5 EXERCISES

Geographical skills key: GS1 Remembering and understanding **GS2** Describing and explaining **GS3** Comparing and contrasting **GS4** Classifying, organising, constructing **GS5** Examining, analysing, interpreting **GS6** Evaluating, predicting, proposing

6.5 Exercise 1: Check your understanding

1. **GS1** What makes a *place* remote?
2. **GS1** How does FIFO reduce remoteness?

For questions 3–5, refer to **FIGURE 6** or download the topographic map of Griffith from the Resources tab.

3. **GS1**
   (a) What is the main use for farmland in the area surrounding Griffith?
   (b) Sketch the symbol of this land use.
   (c) Is this an example of farming type A or type B?
4. **GS3**
   (a) Compare the pattern made by irrigation channels and natural waterways, such as Mirrool Creek.
   (b) Why is irrigation useful in semi-arid areas? At what time of the year do you think it would be mainly used?
   (c) How can you tell from the map that it is not hilly in the areas where there is irrigation farming?
5. **GS5** Identify two natural factors and two human factors that might have influenced people to choose to live in the Griffith area.
6. **GS2** Describe the location of Coober Pedy.
7. **GS2** Why are rural communities under threat?
8. **GS2** Refer to **FIGURE 11**.
   (a) Which regions of Australia are experiencing a decline in population growth?
   (b) Describe where population growth is increasing.

6.5 Exercise 2: Apply your understanding

1. **GS2** Describe the *change* in the speed of settlement of inland Australia that is illustrated by **FIGURE 1**.
2. **GS3** Study **FIGURE 2**. Use your atlas to compare the location of soldier settlements with a rainfall map of Australia. Were soldier settlements located in *places* that receive good rainfall for farming?
3. **GS6** The soldier settlements of 1917 were established on remote, underused land. One hundred years later, would these places still be considered remote? Refer to **FIGURES 2** and **3** in your answer. ▶

---

4. **GS3** Refer to **FIGURE 6**.
    (a) Imagine you travelled in a southerly direction for 2.5 kilometres from Griffith city centre. Now select one square kilometre at this location. Count the number of buildings there are in your chosen square kilometre.
    (b) Continue out from the city edge for at least another 7 kilometres. Choose another square kilometre and count the buildings in your chosen area.
    (c) Compare your results. In which area would you be closer to your neighbours?
    (d) Which one represents **intensive farming**?
5. **GS6** There are many farms in the Griffith region, which means there are many people in the area to support shops, businesses, schools and cultural activities. However, in some parts of Australia, farms are very big and it is a long way to the nearest neighbours. Anna Creek, a beef cattle property in northern South Australia, is 24 000 square kilometres (2 400 000 hectares). The property is in a semi-arid region of South Australia, where vegetation is **sparse** and the nearest town for supplies is 170 kilometres away.
    (a) Use the scale to calculate the number of square kilometres covered by the map in **FIGURE 6**.
    (b) How does this compare to the single farm of Anna Creek?
    (c) At which location, Anna Creek or Griffith, could you most likely satisfy each of the following wishes: to play in a sport team every week, to regularly buy clothes, to collect data about lizards, to grow a lush lawn, to safely learn to drive, to have a private airstrip?

Try these questions in learnON for instant, corrective feedback. Go to www.jacplus.com.au.

6.6 SkillBuilder: Using topographic maps

What are topographic maps?

Topographic maps are a type of map that provides detailed and accurate information of features that appear on the Earth's surface.

They show features of the natural environment, such as forests and lakes, and features of human environments, such as roads and settlements. Relief is often shown using contour lines.

Select your learnON format to access:

- an overview of the skill and its application in Geography (Tell me)
- a video and a step-by-step process to explain the skill (Show me)
- an activity and interactivity for you to practise the skill (Let me do it)
- questions to consolidate your understanding of the skill.

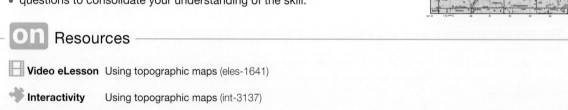

on Resources

🎞 **Video eLesson** Using topographic maps (eles-1641)

🧩 **Interactivity** Using topographic maps (int-3137)

6.7 'Lifestyle' places

6.7.1 Lifestyle

In the years after 1990, the healthy state of the growing Australian and world economies meant that more and more people had jobs and were earning higher incomes. This gave them greater choice as to where and how they wanted to live, and the type of life that they wished to lead, because they had the resources (money) to allow them to choose.

For some people, 'lifestyle choice' means escaping the rush of the modern urban society by choosing a **sea change** or **tree change**. For others, it means using their new-found wealth to fulfil their wants and desires, no matter how wild. Others choose to live in inner-city areas, close to shops, cinemas, restaurants and galleries. Because of this last group of people, governments and businesses have been able convert older industrial areas near the city centre into new activity centres where employment, residences, recreation and services can be found in the one location. Such places are in great demand by those who can afford them, particularly young professionals who want to be near the entertainment and facilities of the inner city.

FIGURE 1 Lifestyle choices for those who have the resources (a) Modern condominiums at Canary Wharf in the centre of London — a lifestyle that has arisen from the old docks (b) The historical apartments above the shops in Mala Strana in the centre of Prague, Czech Republic

FIGURE 2 The contrast between the shantytowns, or favelas, of Rio de Janeiro and their more affluent neighbours is very clear: who has the greater lifestyle choice?

FIGURE 3 Contrasting lives in India (a) The growing middle classes in modern India often purchase modern apartments, like those in this building in New Delhi. (b) For many poor Indians in urban areas, their home and way of life is on the street. Land for housing is expensive in cities such as New Delhi, and beyond the means of many people.

(a)

(b)

FIGURE 4 (a) and (b) The perfect sea change? These houses in the canal district next to Venice Beach, California, give their owners perfect peace and tranquility. (c) ... or do they? (d) The beachfront houses on Venice Beach (e) Condominiums on the cliffs of the Pacific Ocean, in the well-off suburb of Santa Monica, California (f) Many homeless people in California opt to live on Venice Beach because of the climate.

6.7.2 CASE STUDY: Fishermans Bend urban renewal project

Fishermans Bend is an urban renewal project located within five kilometres of Melbourne's CBD. With a focus on environmental, economic and social sustainability, Fishermans Bend will support the growth of Melbourne by accommodating 80 000 residents and 80 000 jobs by 2050.

Key elements of the planning controls include:
- the introduction of a Dwelling Density Ratio (a mix of high-, medium- and low-rise buildings)
- social housing for low-income people
- overshadowing controls to protect public open space

- the encouragement of dwelling diversity and a range of building types
- water storage and reuse across buildings
- a requirement for all new buildings to meet a minimum 4 Star Green Star rating and large-scale buildings to meet a 5 Star Green Star rating
- delivery of 6 per cent affordable housing
- delivery of public open space, major roads and community infrastructure
- minimum employment floor space in designated core areas.

FIGURE 5 The Melbourne 2050 Plan incorporates the Fishermans Bend urban renewal area.

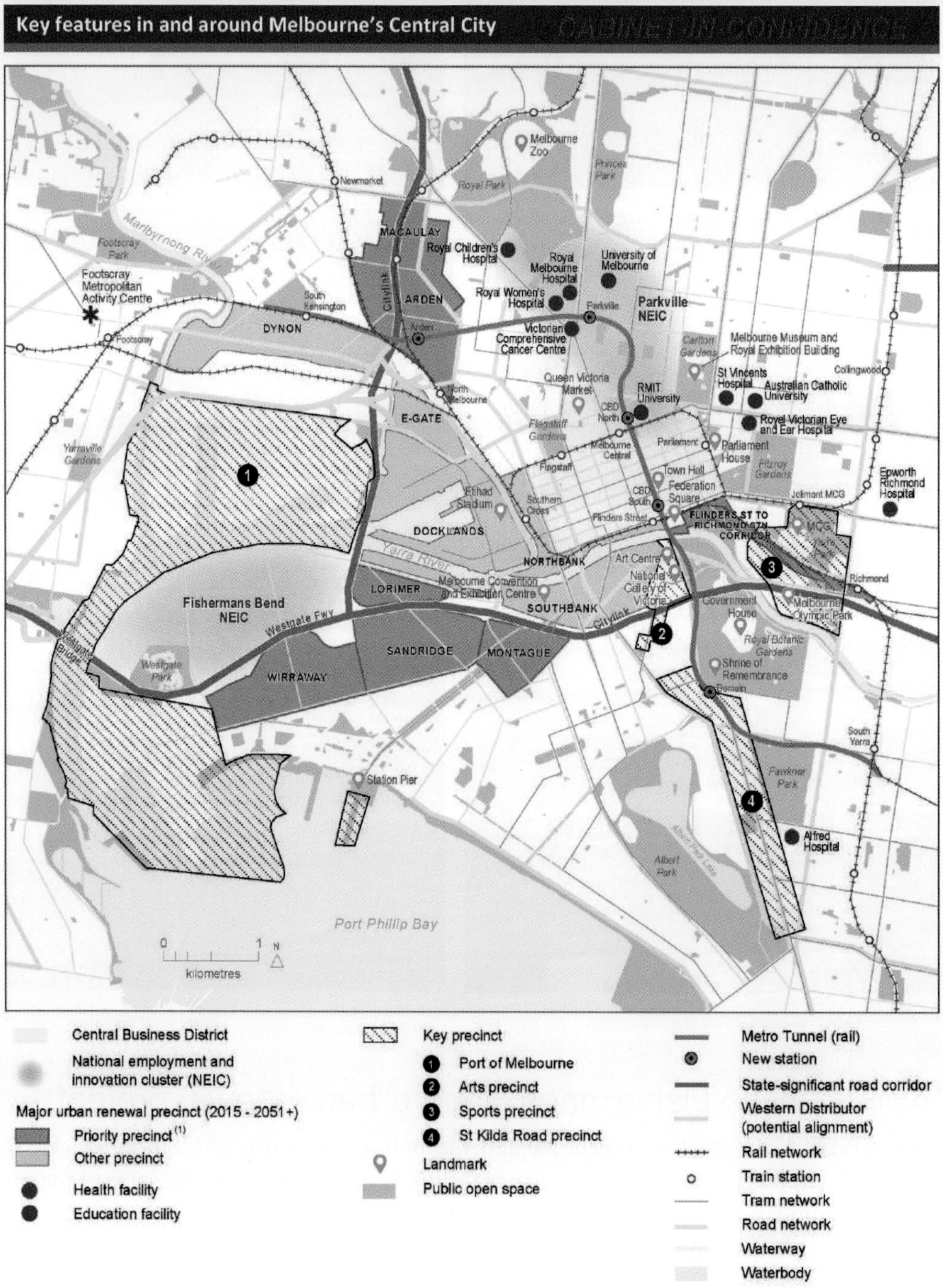

Key features in and around Melbourne's Central City

Legend:

Central Business District

National employment and innovation cluster (NEIC)

Major urban renewal precinct (2015 - 2051 +)
- Priority precinct [1]
- Other precinct
- Health facility
- Education facility

Key precinct
1. Port of Melbourne
2. Arts precinct
3. Sports precinct
4. St Kilda Road precinct

Landmark

Public open space

Metro Tunnel (rail)

New station

State-significant road corridor

Western Distributor (potential alignment)

Rail network

Train station

Tram network

Road network

Waterway

Waterbody

FIGURE 6 Existing and proposed cycling infrastructure and open space at Fishermans Bend urban renewal area

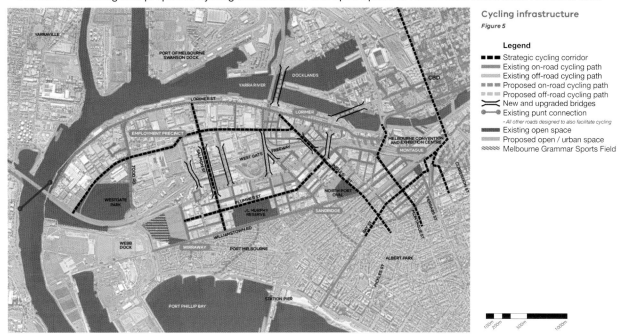

Cycling infrastructure
Figure 5

Legend
- ▬▬▬ Strategic cycling corridor
- ▭▭▭▭ Existing on-road cycling path
- ▭▭▭▭ Existing off-road cycling path
- ▬ ▬ ▬ Proposed on-road cycling path
- ▭ ▭ ▭ Proposed off-road cycling path
- ⋈ New and upgraded bridges
- ●— Existing punt connection
- · *All other roads designed to also facilitate cycling*
- ▭▭▭▭ Existing open space
- ▭▭▭▭ Proposed open / urban space
- ▨▨▨▨ Melbourne Grammar Sports Field

on Resources

🔎 **Google Earth** Canary Wharf, London

Prague, Czech Republic

Rio de Janeiro

Fishermans Bend

6.7 INQUIRY ACTIVITIES

1. In groups of four, use a graphics software program to create a concept wheel that explores the meaning of the word 'lifestyle'. Display all the wheels in a class presentation.
   As a class, compare and discuss the lifestyle wheels that you produced. Is there a common view of lifestyle that is representative of the whole class?   **Classifying, organising, constructing**
2. Use Google Earth to zoom in on Fishermans Bend. Can you identify any of the areas and planning described in the text or on the map? Use Google Earth and other online resources to discover how the development of this area is proceeding.   **Examining, analysing, interpreting**

6.7 EXERCISES

Geographical skills key: GS1 Remembering and understanding **GS2** Describing and explaining **GS3** Comparing and contrasting **GS4** Classifying, organising, constructing **GS5** Examining, analysing, interpreting **GS6** Evaluating, predicting, proposing

6.7 Exercise 1: Check your understanding

1. **GS1** What is meant by the term 'lifestyle choice'?
2. **GS5** Study **FIGURES 1, 2, 3** and **4**. Identify the features in the photographs that indicate that the people living in these *places* have the resources to choose to live in such *environments*.
3. **GS3** Outline the differences between a sea *change* and a tree *change*?   ▶

4. **GS3** Compare the images in **FIGURES 2** and **3**. What similarities and differences can you see in the 'lifestyle' choices of these residents of Rio de Janeiro and New Delhi? How much real power do they have to decide where they live?
5. **GS2** Refer to **FIGURE 5**.
   (a) Describe the location of Fishermans Bend in relation to Melbourne's CBD.
   (b) What features and infrastructure might attract people to live here?
6. **GS2** Refer to **FIGURE 6**. Describe the existing and planned options for cycling and open space at Fishermans Bend. How will these improve liveability for the residents?

6.7 Exercise 2: Apply your understanding

1. **GS6** Would you like to live at Fishermans Bend? List what you would consider positive and negative aspects. Ask your family about how they would feel about living there.
2. **GS6** Study the lifestyle *environment* in **FIGURE 4**. From the evidence in the images, identify and list any possible *changes* (natural or human) that might affect the liveability of Venice Beach and Santa Monica.
3. **GS6** Would you like to live in a *place* like Venice Beach? Give reasons for your answer.
4. **GS2** Name two *tree change* locations that you or your family would like to live in and why.
5. **GS2** Name two *sea change* locations that you or your family would like to live in and why.

Try these questions in learnON for instant, corrective feedback. Go to www.jacplus.com.au.

6.8 Liveable places in Australia

6.8.1 My place

What is your **neighbourhood** or local place like? All of us live in a community, and these are often centred on the place where we live, go to school or work.

Teenagers have different types of local places that have special meaning for them, each one at a different scale: their bedroom, home and neighbourhood.

When you live in a neighbourhood, you become familiar with all the things that help to create the character of the place. Sometimes a neighbourhood is made up of people who have similar interests and beliefs, whether these be cultural, sporting, environmental or job-related. Other neighbourhoods have a mixture of people from different backgrounds, creating a vibrant, multicultural community identity. The fact that Australian neighbourhoods can be so different is what makes Australia such an interesting place to live in.

Neighbourhoods have always existed in Australia. The 'country' that is special to the many Aboriginal and Torres Strait Islander nations is often based on language. For instance, the Kulin Nation consists of five Victorian Aboriginal communities who lived in what is now the Melbourne region before Europeans invaded. Each community spoke its own language and controlled a region that had definite boundaries (see **FIGURE 1**). Within each community, there were different dialects that overlapped. These dialects were spoken by different clans — groups of related families. Thus, these nations saw, and often still see, their neighbourhood as the region in which people spoke the same language and had the same customs, such as marriage rituals. People were, and are, socially connected.

Because nearly 90 per cent of Australians live in towns and cities, most people are likely to live in a street that is part of a suburb, town or city, and which itself is part of a state or territory. On the other hand, there are many Australians who do not live in urban areas, but still live in their own communities that are just as distinctive as neighbourhoods in towns and cities. How can we describe where our local place is and what it is like? Sometimes, people try to use words to do this, but it is not an easy task. Geographers have no such trouble, however; they can use maps.

FIGURE 1 The places belonging to Indigenous Australian peoples

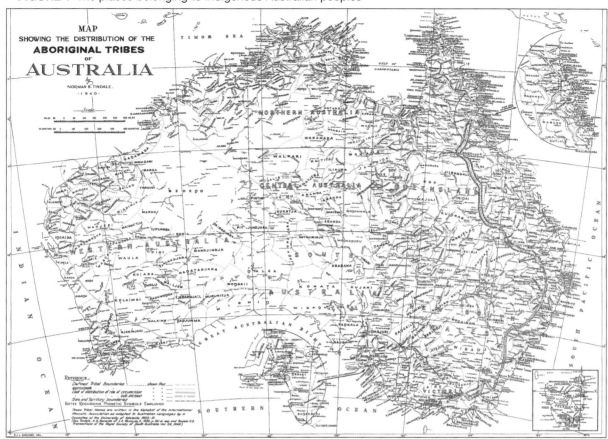

Map Showing the Distribution of the Aboriginal Tribes of Australia, Norman B. Tindale, 1940. Users of this map should be aware that certain words, terms or descriptions may be culturally sensitive and may be considered inappropriate today, but may have reflected the creator's attitude or the period in which they were written. Borders and terminology used may be contested in contemporary contexts.

FIGURE 2 **Mental map** of Jayden's local place (a) by Jayden and (b) by Annette, Jayden's mother

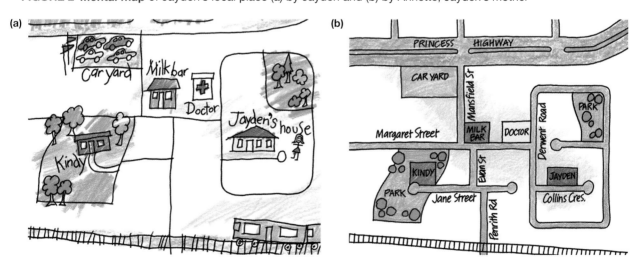

6.8.2 What makes Australia so liveable?

Where is your favourite place in Australia? Have you been to a holiday paradise, one that you think would be the perfect place to live? Is the climate perfect, the scenery spectacular? Is it safe, fun and the place for adventure? Is this place in a city, in the **wilderness** or in the next street? Is it paradise because your friends or family live there or because of the natural or **built environment**?

Among the most popular and beautiful tourist destinations in Australia are the Great Barrier Reef, Uluru, Melbourne, Sydney, the Gold Coast, the Great Ocean Road, Monkey Mia, Kakadu, the Tasmanian Wilderness, the Blue Mountains, Port Arthur, Byron Bay, Kangaroo Island and Ningaloo Reef. Many of these places have unique landscapes, located within naturally stunning environments. Four of these are predominantly built environments: Sydney, Melbourne, the Gold Coast and Port Arthur. The remaining ten places are best known for their natural, often remote, and almost wilderness environments.

Some of these wonderful places are found in or close to cities and large towns; some have significant local populations; and some are quite remote. They are all places that attract large numbers of visitors every year. People come to see or experience an aspect of the local environment that brings them pleasure. These places are often perfect for a holiday but they may also be a place to live. Is it mostly the excitement of a big city, natural beauty, or some other factor that makes you decide which place is the most liveable?

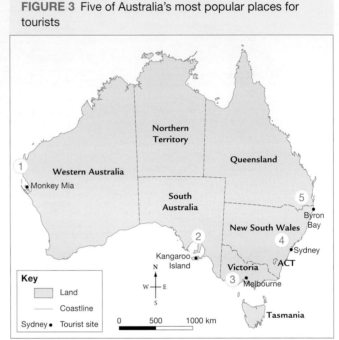

FIGURE 3 Five of Australia's most popular places for tourists

Source: Spatial Vision

1. **Monkey Mia** is an environment where you can experience natural wildlife by interacting with dolphins. Monkey Mia is located in Shark Bay on the coast of Western Australia, 850 kilometres north of Perth. For over 40 years, a small pod of dolphins has come ashore to connect with beachgoers. The Department of Environment and Conservation provides staff who supervise the feeding of fish to these dolphins each day. It is an unusual opportunity for people to see wild dolphins up close, quite near to the shore. Monkey Mia is a place of great natural beauty without a huge tourist resort attached. Most visitors camp. It is an important stop on the around-Australia tourist trail. Fewer than 800 residents currently live near the Monkey Mia Resort.

2. **Kangaroo Island** is a place of natural beauty. It is Australia's third largest island, found about 160 kilometres south of Adelaide. It is a wildlife lover's paradise, being home to many native Australian animals in their natural habitats, including koalas, kangaroos, seals and penguins. It has remote, unspoiled beaches and interesting rocky outcrops. Although first settled in the late 1830s, its present population of over 4200 is the highest it has ever been. It was originally settled as a fishing and farming community but today is better known as a tourist destination.

3. **Melbourne** is the second-largest and most **liveable city** in Australia (2011–2017; second to Vienna in 2018 according to the *Economist* magazine). It is the capital of Victoria and home to over five million residents in 2018. It is an attractive destination for tourists, who enjoy visiting its major sporting and cultural events, shops, restaurants and theatres. Melbourne is located beside Port Phillip Bay and on the Yarra River. It is not a city known for its beautiful natural environment, but it has become known for its distinctive laneways, bars and café culture.

4. **Sydney** is a built environment in a beautiful setting and is Australia's largest and oldest city. It is often called the 'Harbour City'. Sydney is popular with both domestic and international tourists and is home to nearly 5.5 million residents in 2018. It has many attractions, including restaurants, beautiful beaches, theatres, galleries and iconic landmarks. It has a beautiful natural environment with varied experiences provided by the built environment. This makes it an extremely popular destination for everyone.

5. **Byron Bay** is a beachside town in northern New South Wales, located 160 kilometres south of Brisbane. Byron Bay is a very relaxed place with a local community that includes many artists and retired hippies. It is an important surfing place, with easy access to offshore reefs and stunning beaches. It has become a popular place for 'schoolies' end-of-year celebrations. Byron Bay Shire had a population of nearly 34 000 people in 2017 (9500 in the township of Byron Bay), who rely heavily on tourism and agriculture for their income.

6.8 INQUIRY ACTIVITY

Create a mental map of your neighbourhood or local *place*. Locate your house in the centre of the sheet and work outwards from there. The map should be as detailed as possible. Include features such as:

- streets and their names
- houses of friends or family
- shops, parks, trees, post boxes, telephone poles, pedestrian crossings, railway lines and stations
- anything you can remember, but the map must be drawn from memory.

Present the map using geographical rules (BOLTSS). Since you are not drawing the map to a *scale*, write 'Not to scale' in the correct position. Remember to use conventional colours and symbols as far as possible. Compare your mental map to an actual map of your neighbourhood.

a. In what ways was your map accurate?
b. Which features did you not mark on your map?
c. Which parts of your neighbourhood did you know well and which did you not know well?
d. Think of reasons to explain your answers to (c).

Design a map of your most liveable *place*. Consider the natural and built *environments*; distance to a city, services, job and recreational opportunities; climate; and lifestyle. Annotate your map to explain why this is where you would like to live. Use the **Nothing like Australia** weblink in the Resources tab to help find your ideal location.

6.8 EXERCISES

Geographical skills key: GS1 Remembering and understanding **GS2** Describing and explaining **GS3** Comparing and contrasting **GS4** Classifying, organising, constructing **GS5** Examining, analysing, interpreting **GS6** Evaluating, predicting, proposing

6.8 Exercise 1: Check your understanding

1. **GS1** What is a *neighborhood*?
2. **GS1** What percentage of Australians live in cities and towns?
3. **GS1** In your own words, define the term 'liveable city'.
4. **GS4** Classify each of the following places in Australia as a natural environment, built environment, or both.

Monkey Mia	Melbourne
Kakadu	Great Barrier Reef
Sydney	The Gold Coast
The Blue Mountains	Port Arthur

5. **GS1** Which city was named the 'most liveable city' in the world in 2018?

6.8 Exercise 2: Apply your understanding

1. **GS2** Brainstorm a list of the features to describe your most liveable *place*.
2. **GS3** Study **FIGURE 2**, which shows two mental maps of the same neighbourhood *place*. One is drawn by Jayden, a seven-year-old boy, and the other is drawn by his mother.
   (a) Compare the two maps by drawing up a table like the one below and filling in the spaces. ▶

	Features that are different	Features that are similar
Land use		
Transport		
Street layout		
Relative sizes		
Names of places		
Other		

(b) Suggest reasons to explain the major similarities and differences between the maps drawn by Jayden and his mother. Think about factors such as age, duties during the day, transport and friendships.

3. **GS6** Is your most liveable *place* in a natural or a built *environment* or a mixture of the two?

4. **GS5** After reading the paragraphs numbered 1, 2, 3, 4 and 5, which of these *places* is most similar to your most liveable *place*? Explain your answer.

5. **GS6** If you wished to work as a national park ranger, which of the *places* in **FIGURE 3** would be best and why?

6. **GS6** If you were planning a career in the theatre, which of the *places* in **FIGURE 3** would be best and why?

7. **GS6** If you wished to live in a relaxed coastal *environment* close to a capital city, which of the *places* in **FIGURE 3** would be best and why?

Try these questions in learnON for instant, corrective feedback. Go to www.jacplus.com.au.

6.9 SkillBuilder: Creating a concept diagram

on**line** only

What is a concept diagram

A concept diagram, sometimes mistakenly called a concept map, is a graphical tool that shows links between ideas, or concepts. Concept diagrams organise links into different levels.

Concept diagrams enable you to organise your ideas and communicate them to others.

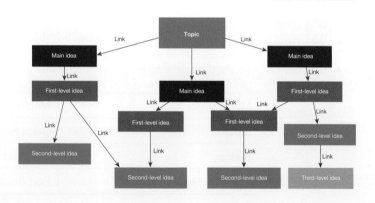

Select your learnON format to access:

- an overview of the skill and its application in Geography (Tell me)
- a video and a step-by-step process to explain the skill (Show me)
- an activity and interactivity for you to practise the skill (Let me do it)
- questions to consolidate your understanding of the skill.

 Resources

Video eLesson Creating a concept diagram (eles-1640)

Interactivity Creating a concept diagram (int-3136)

6.10 Thinking Big research project: Harmony Week presentation

SCENARIO

Harmony Week is celebrated each year in Australia and celebrates Australia's multicultural diversity. You have been asked to prepare a special Harmony Week presentation at your school that focuses on the importance of place and Country to Aboriginal Australians.

Select your learnON format to access:
- the full project scenario
- details of the project task
- resources to guide your project work
- an assessment rubric.

 Resources

projectsPLUS Thinking Big research project: Harmony Week presentation (pro-0237)

6.11 Review

6.11.1 Key knowledge summary
Use this dot point summary to review the content covered in this topic.

6.11.2 Reflection
Reflect on your learning using the activities and resources provided.

 Resources

 eWorkbook   Reflection (doc-32139)

Crossword (doc-32140)

 Interactivity   Choosing a place to live crossword (int-7703)

KEY TERMS

arid lacking moisture; especially having insufficient rainfall to support trees or plants

built environment a place that has been constructed or created by people

community a group of people who live and work together, and generally share similar values; a group of people living in a particular region

fly in, fly out (FIFO) describes workers who fly to work in remote places, work 4-, 8- or 12-day shifts and then fly home

horticulture the growing of garden crops such as fruit, vegetables, herbs and nuts

intensive farming farming that uses a lot of resources per hectare and changes the look of the region

irrigation water provided to crops and orchards by hoses, channels, sprays or drip systems in order to supplement rainfall

liveable city a city that people want to live in, which is safe, well planned and prosperous and has a healthy environment

location a point on the surface of the Earth where something is to be found

mental map a drawing or map that contains our memory of the layout and distribution of features in a place

neighbourhood a region in which people live together in a community

pastoral describes land used for keeping, or grazing, sheep or cattle

place specific area of the Earth's surface that has been given meaning by people

pull factors positive aspects of a place; reasons that attract people to come and live in a place

push factors reasons that encourage people to leave a place and go somewhere else

region any area of varying size that has one or more characteristics in common

remote describes a place that is distant from major population centres

sea change the act of leaving a fast-paced urban life for a more relaxing lifestyle in a small coastal town

sparse thinly scattered or unevenly distributed; often used when referring to population density

tree change the act of leaving a fast-paced urban life for a more relaxing lifestyle in a small country town, in the bush, or on the land as a farmer

urban decay situation in which a city area has fallen into a state of disrepair through its people leaving the area or not having enough resources to look after it

wilderness a natural place that has been almost untouched or unchanged by the actions of people

7 A world of people and places

7.1 Overview

Rural, urban, remote or central, the world is full of places. Are any two places the same?

7.1.1 Introduction

We all live in different places. Places are important to people, whether they are rural or urban, remote or central, permanent or temporary. But no two places are alike; they differ in aspects such as their appearance, size and features. In your mind's eye, try to picture the similarities and the differences between places such as a country town, a popular tourist destination, a remote village overseas, a scientific base in Antarctica, an Australian Indigenous community and a mining town. You may think of others to add to this list.

Resources

 eWorkbook   Customisable worksheets for this topic

Video eLesson Why we live where we do (eles-1620)

LEARNING SEQUENCE

To access a pre-test and starter questions and receive immediate, **corrective feedback** and **sample responses** to every question select your learnON format at www.jacplus.com.au.

7.2 Why we live where we do

7.2.1 Historic settlements and transport networks

Does the availability of rainfall explain why Australians currently live where they do? Could it also be warmer weather, good soils, and access to the coast, mineral resources, people and flat land?

Australia's unique physical characteristics have led to some interesting patterns in the development of our cities and urban areas. As we know, deserts take up over 70 per cent of Australia. Without the resources needed to sustain large populations, the majority of land in Australia remains sparsely inhabited. Instead, 85 per cent of Australians live within 50 kilometres of the coast.

Unlike Indigenous Australian peoples, the people who occupied Australia after 1788 have not been tied to a particular place in Australia by their understanding of '**country**'. The liveability of places can be influenced by their access to resources. Indigenous people continue to live in many places across Australia with their nomadic lifestyles and sustainable ways of living allowing habitation in a range of environments.

Australia's first large European colonies were all built close to rivers and harbours, which provided safe anchorage for their sailing ships. The early colonisers and convicts, along with their possessions and food, were all transported from Europe by ship. These colonisers were quite wary of the unknown inland of the continent, so they clung to the coast and relied on sea transport. The sea allowed goods to be imported and exported; it provided fishing, whaling and sealing; and it brought cooler weather due to onshore breezes. Rivers, such as the Yarra (seen in **FIGURE 1**), provided water for household, industrial and agricultural use, as well as a safe port for passengers and cargo.

FIGURE 1 The Yarra River, Melbourne, 1864

Sydney, Melbourne, Brisbane, Hobart, Adelaide and Perth were generally the first settlements in each colony. They have maintained their importance as their states' largest cities and their centres of government, transport and commerce. Big cities are great places to live because they provide many opportunities for work, education, healthcare and recreation.

Land use

Beginning in 1788, Europeans and subsequent migrants have changed the natural environment by clearing vegetation and using the land for building cities and farms and creating the services people need.

Examine **FIGURE 2** and try to work out the difference between **intensive** and **extensive land use** regions. The intensive regions are mostly located closer to the coast than extensive regions, and they are smaller areas. Compare this land use map to a rainfall map in your atlas. Does more rain tend to fall in the intensive or the extensive land use regions?

In places where rainfall is low or unreliable, such as central Australia, grass growth is seasonal. This semi-arid interior does not support the variety of land uses that are possible closer to the coast.

FIGURE 2 Major land use regions of Australia

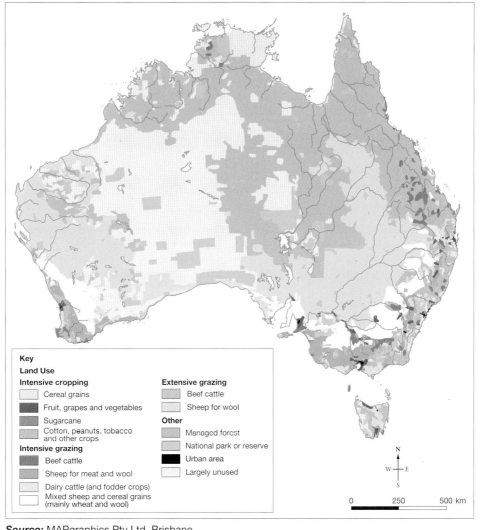

Key

Land Use

Intensive cropping
- Cereal grains
- Fruit, grapes and vegetables
- Sugarcane
- Cotton, peanuts, tobacco and other crops

Intensive grazing
- Beef cattle
- Sheep for meat and wool
- Dairy cattle (and fodder crops)
- Mixed sheep and cereal grains (mainly wheat and wool)

Extensive grazing
- Beef cattle
- Sheep for wool

Other
- Managed forest
- National park or reserve
- Urban area
- Largely unused

0   250   500 km

Source: MAPgraphics Pty Ltd, Brisbane

Jobs

People need to work in order to provide for their basic needs. **FIGURE 3** shows that the majority of jobs in 2016 were in the sectors of services (health, finance, education and administration), retail trade and construction. Most of these job opportunities are available in major cities and regional centres, which are generally close to the coast and in areas of higher rainfall.

There are few job opportunities in dry and remote regions of Australia. Agriculture and mining provide some employment in these places. Some jobs are also available in manufacturing (processing the output of farms and mines) and in services, such as healthcare and tourism. Mining jobs in remote places are declining due to the reduced international price for minerals and mine closures and very few new mines opening. Remote mining sites have skills shortages that require employers to pay very high wages to attract workers.

FIGURE 3 Australia's job sectors, 2016

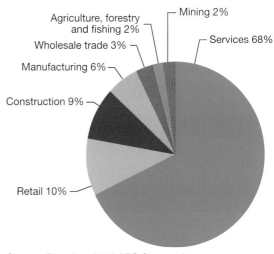

- Agriculture, forestry and fishing 2%
- Mining 2%
- Wholesale trade 3%
- Services 68%
- Manufacturing 6%
- Construction 9%
- Retail 10%

Source: Based on 2016 ABS Census data

7.2 EXERCISES

Geographical skills key: GS1 Remembering and understanding **GS2** Describing and explaining **GS3** Comparing and contrasting **GS4** Classifying, organising, constructing **GS5** Examining, analysing, interpreting **GS6** Evaluating, predicting, proposing

7.2 Exercise 1: Check your understanding

1. **GS1** In your own words, define the term 'country'.
2. **GS1** What percentage of Australians live within 50 kilometres of the coast?
3. **GS1** List four reasons why the state capital cities were settled.
4. **GS1** What is the largest job sector in Australia?
5. **GS1** In Australia, are more workers employed in mining or in manufacturing?

7.2 Exercise 2: Apply your understanding

1. **GS3** What is the difference between intensive and extensive land use? Give examples of each type.
2. **GS2** What trends can you see in the **FIGURE 2** map? Where are the regions in which intensive agriculture occurs?
3. **GS6** 'Access to water improves the liveability of a place.' Write a paragraph that agrees with this statement and another that disagrees with this statement.
4. **GS6** A doctor may be offered very high wages to work in a remote, dry place in north-western Australia. What economic, *environmental* and social measures might they use to decide between the liveability of this place and their current workplace in Melbourne?
5. **GS2** Only a small amount of Australia's total land area is used for urban development. Give a reason for this.

Try these questions in learnON for instant, corrective feedback. Go to www.jacplus.com.au.

7.3 SkillBuilder: Understanding satellite images

What are satellite images?

Satellite images are images that show parts of our planet from satellites in space and transmitted to stations on Earth. Satellite images help geographers observe a much larger area of the Earth's surface than photographs taken from an aircraft.

Select your learnON format to access:

- an overview of the skill and its application in Geography (Tell me)
- a video and a step-by-step process to explain the skill (Show me)
- an activity and interactivity for you to practise the skill (Let me do it)
- questions to consolidate your understanding of the skill.

Resources

📽 **Video eLesson** Using alphanumeric grid references (eles-1643)

🧩 **Interactivity** Using alphanumeric grid references (int-3139)

7.4 Growth cities in Australia

7.4.1 The growth suburbs

Which place in Australia is growing the fastest? If a place is liked by lots of people, does that make it the best? What makes a suburb the most popular? Coastal areas have always been a popular place for Australians to relax and holiday. Is the fastest growing place in Australia near the coast?

People might move to a new place for many different reasons. The attractions that entice people to live somewhere are called its **pull factors**. Pull factors include cheaper housing, better climate, more job opportunities and improved lifestyle. People can also be forced to leave their home and move to a new place. These reasons are known as **push factors**. Loss of your job or business, poor school or health facilities, and a natural disaster, such as flood or fire, are examples of push factors.

The Australian Bureau of Statistics, which collects information for the Australian Government, says that three of the five fastest growing suburbs are in Melbourne. The other two suburbs are in Canberra and Brisbane respectively. Over recent years, Melbourne has been Australia's fastest growing capital city. It is not surprising that it often tops tables as the world's most liveable city.

What makes the Gold Coast such a great place to live?

The Gold Coast's warm weather, beach culture and holiday lifestyle have attracted many new residents. **FIGURE 1** shows that most of the new arrivals came from New South Wales. Many were attracted to the place their family visited on holiday, and they later decided to make it their permanent home. The Gold Coast is now the sixth-largest urban area in Australia. It is a major tourist destination, offering a wide range of work opportunities, community facilities and intercity and interstate transport links by road, rail and air. Many new residents are older Australians who have retired to this place.

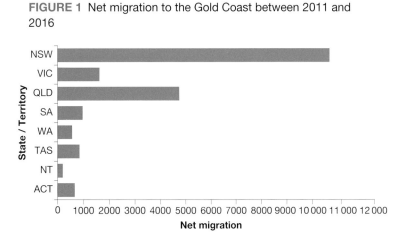

FIGURE 1 Net migration to the Gold Coast between 2011 and 2016

The increased population has placed pressure on the coastal environment, as well as on the existing infrastructure of schools, hospitals, roads and housing.

Use Nearmap or Google Maps to access an **aerial photograph** of the Gold Coast region today. Compare this to **FIGURE 2**, showing the changes that have occurred to the land use here over the past 50 years.

Why was the Gold Coast chosen for an AFL expansion team and the 2018 Commonwealth Games?

The main AFL states are Victoria, South Australia, Western Australia and Tasmania along with the Indigenous communities of the Northern Territory. In March 2009, the Gold Coast Football Club, now named the Gold Coast Suns, was established, supported financially by the Australia Football League (AFL). The club's establishment on the Gold Coast has seen a rise in youth participation in AFL.

The six largest Australian cities now have at least one AFL club: Sydney (2 clubs), Melbourne (9 clubs), Brisbane (1 club), Perth (2 clubs), Adelaide (2 clubs) and Gold Coast (1 club).

The Gold Coast was chosen to host the Commonwealth Games in 2018. Metricon Stadium at Carrara, the home of the Gold Coast Suns, was temporarily transformed and increased in capacity to host the athletics events as well as the opening and closing ceremonies for the Commonwealth Games. Australia has hosted five Commonwealth Games, but this is the first time that they have not been held in a state capital city.

FIGURE 2 Topographic map extract of the Gold Coast region in 1967

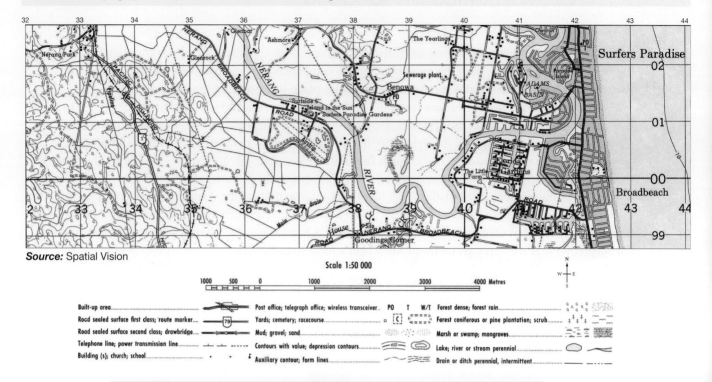

Source: Spatial Vision

Scale 1:50 000

Built-up area....................		Post office; telegraph office; wireless transceiver.. PO T W/T		Forest dense; forest rain................	
Road sealed surface first class; route marker.....	79	Yards; cemetery; racecourse................... □ [C]		Forest coniferous or pine plantation; scrub..........	
Road sealed surface second class; drawbridge....		Mud; gravel; sand...................		Marsh or swamp; mangroves................	
Telephone line; power transmission line............		Contours with value; depression contours............		Lake; river or stream perennial................	
Building (s); church; school................		Auxiliary contour; form lines................		Drain or ditch perennial, intermittent............	

FIGURE 3 The ground developed for the Gold Coast Suns AFL team

on Resources

Digital document Topographic map extract — Gold Coast region (doc-11400)

7.4 INQUIRY ACTIVITIES

1. Compare **FIGURE 2** with the Nearmap or Google Maps aerial image of the same *place* today.
   (a) Study the map and photo in small groups. Identify the changes to the *environment*, both built and natural, between the map and the photo.
   (b) Collate this information in a table.
   (c) Write one sentence to describe the change to the built *environment*.
   (d) Write one sentence to describe the change to the natural or physical *environment*.
   (e) The population of the Gold Coast is predicted to double to 1.2 million people by 2050. Is there much *space* left in this area of the Gold Coast for housing? Suggest where new suburbs could be established.

Examining, analysing, interpreting

2. Using your knowledge of the factors that have made the Gold Coast so liveable and grow so fast, use your atlas to identify another *place* that could become the 'new' Gold Coast. Identify the pull factors for your location and write a paragraph to sell its advantages to potential residents. **Evaluating, predicting, proposing**

7.4 EXERCISES

Geographical skills key: GS1 Remembering and understanding **GS2** Describing and explaining **GS3** Comparing and contrasting **GS4** Classifying, organising, constructing **GS5** Examining, analysing, interpreting **GS6** Evaluating, predicting, proposing

7.4 Exercise 1: Check your understanding

1. **GS1** What features of the Gold Coast have made it grow so quickly?
2. **GS1** Where have most of the new residents of the Gold Coast come from?
3. **GS2** What services, facilities and environmental attractions does the Gold Coast offer to people wishing to find a more liveable *place* to retire to?
4. **GS1** What are *push factors*? Give an example.
5. **GS1** What are *pull factors*? Give an example.
6. **GS2** What was significant about the Gold Coast hosting the 2018 Commonwealth Games?

7.4 Exercise 2: Apply your understanding

1. **GS2** Use examples to explain the difference between the environment, services and facilities that make up the push and pull factors which result in the rise or fall of the population of a place.
2. **GS6** Think about the *place* where you live. Make a list of the pull factors that make your town or suburb more liveable. Then list the possible push factors that might make someone leave your suburb or town to live somewhere else.
3. **GS6** Imagine you are a town planner. Suggest two new features that you can add to your suburb which would make it a more appealing place to live.
4. **GS6** Many of Australia's towns and cities are growing faster than ever before. What do you believe is behind this growth?
5. **GS3** Consider the push and pull factors that you've identified and discussed already. Which do you believe is the most important factor people consider when deciding where to live? Justify your response.

Try these questions in learnON for instant, corrective feedback. Go to www.jacplus.com.au.

7.5 SkillBuilder: Using alphanumeric grid references

What are alphanumeric grid references?

Alphanumeric grid references are a combination of letters and numbers that help us locate specific positions on a map. Letters and numbers are placed alongside the gridlines, just outside a map. The grid, letters and numbers allow you to pinpoint a place or feature by stating its alphanumeric grid reference.

Select your learnON format to access:

- an overview of the skill and its application in Geography (Tell me)
- a video and a step-by-step process to explain the skill (Show me)
- an activity and interactivity for you to practise the skill (Let me do it)
- questions to consolidate your understanding of the skill.

Resources

Video eLesson Using alphanumeric grid references (eles-1642)

Interactivity Using alphanumeric grid references (int-3138)

7.6 Life in a country town

7.6.1 The attraction of the country

Country towns come in all shapes and sizes. They can be small centres with a post office and general store or they can be substantial towns. Because most of Australia's population and businesses are concentrated in the capital cities, even people who live in quite large towns outside the capital cities see themselves as living in the country.

Even though most Australians live in large urban centres, the rural or country regions are very important because this is where food is grown, **natural resources** are extracted and ecosystems can flourish. Many Australians travel to country places for holidays and many dream of moving to the country. The attractions of country places include cheaper housing, less traffic, a greater sense of safety, and the allure of living within and around natural environments.

COUNTRY LIFE

How do you feel about living in different places? The following nine statements refer to different opinions about living in rural areas or cities.
- Rural areas are peaceful, have lots of space and clean air.
- Cities provide more choice in activities and places to live.
- I feel isolated in cities.
- Pollution and noise in big cities impair living conditions.
- I don't feel safe in big cities because of crime.
- Rural areas have great communities with people supporting one another.
- The natural environment in many rural areas is very attractive.
- I feel isolated in rural areas.
- Jobs and transport are more accessible in cities.

Complete a diamond ranking diagram by writing the statement you most agree with at the top and the one you most disagree with at the bottom. Then choose the next two top and bottom statements and the final three in the middle of the ranking.

Explain your ranking to another person. How might these rankings change if they were completed by people who live in places different from where you live? Can you test this hypothesis?   **[Intercultural Capability]**

7.6.2 Demography

The **demographic** characteristics of country places are influenced by location and activities in the surrounding area.

For instance, Leongatha is located on the South Gippsland Highway, 135 kilometres south-east of Melbourne, Victoria. Reliable rainfall and good soil make the area one of the most productive in Victoria. Dairy farming is the main type of farming, and the milk-processing factory is the largest single employer in town.

Another town, Coleraine, is located on the Glenelg Highway, 350 kilometres west of Melbourne. The farms are generally large. Sheep and cattle grazing are the main types of farming, and there is no major business in the town.

TABLE 1 shows the current and predicted population demographics for Melbourne, Leongatha and Coleraine.

TABLE 1 Predicted population for selected Victorian places, 2031

Local government area	2011		2031	
Municipality	% aged under 20	% aged over 65	% aged under 20	% aged over 65
Melbourne (urban)	24.3	23.8	13.0	17.1
South Gippsland Shire (rural includes Leongatha)	24.8	21.9	19.5	28.1
Southern Grampians Shire (rural includes Coleraine)	25.8	21.6	19.6	30.3

A sense of belonging

In the country there are opportunities to be involved in a wide range of community activities. Activities might have an economic focus (such as fundraising), an environmental focus (such as Landcare) or a social focus (such as youth groups). A common outcome of all activities is the way they contribute to a sense of connectedness or belonging.

FIGURE 1 Growing up in a country town might mean …

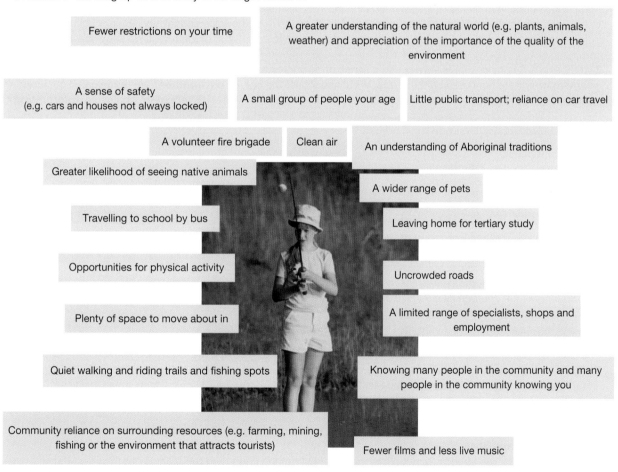

Fewer restrictions on your time

A greater understanding of the natural world (e.g. plants, animals, weather) and appreciation of the importance of the quality of the environment

A sense of safety (e.g. cars and houses not always locked)

A small group of people your age

Little public transport; reliance on car travel

A volunteer fire brigade

Clean air

An understanding of Aboriginal traditions

Greater likelihood of seeing native animals

A wider range of pets

Travelling to school by bus

Leaving home for tertiary study

Opportunities for physical activity

Uncrowded roads

Plenty of space to move about in

A limited range of specialists, shops and employment

Quiet walking and riding trails and fishing spots

Knowing many people in the community and many people in the community knowing you

Community reliance on surrounding resources (e.g. farming, mining, fishing or the environment that attracts tourists)

Fewer films and less live music

DISCUSS

As a class or group, discuss the meaning of the term 'country life'. Create a list of words that describe your agreed view.

Describing and explaining

It is common for even small towns to provide a range of sports. Sports provided in a town as large as Leongatha might include Australian Rules football, cricket, little athletics, tennis, equestrian events, bowls, fishing, cycling, croquet, skateboarding, golf, swimming, basketball, netball, table tennis, badminton, karate, gymnastics, squash and taekwondo. Also available are cultural activities and entertainment, such as films, brass bands, guides, scouts, art galleries, dancing and theatre groups.

FIGURE 2 Lawn bowls is a popular sport in small towns.

on Resources

🧩 **Interactivity**   Country town services (int-3092)

🌏 **Google Earth**   Leongatha

7.6 INQUIRY ACTIVITY

1. (a) Create a list of Australian country towns that you have heard of.
   (b) Collaborate with others to map the location of each of these towns.
   (c) For each town, record its latitude, distance and direction from the state capital city.
   (d) Complete the following generalisation about the known towns: 'Most of the towns we know are located
   _____.'   **Remembering and understanding**
2. Indigenous Australians have always had a sense of belonging to a *place*, and elements of the natural *environment* are honoured on their flags. What aspects of the natural *environment* are represented on each of these flags?   **Examining, analysing, interpreting**

7.6 EXERCISES

Geographical skills key: GS1 Remembering and understanding **GS2** Describing and explaining **GS3** Comparing and contrasting **GS4** Classifying, organising, constructing **GS5** Examining, analysing, interpreting **GS6** Evaluating, predicting, proposing

7.6 Exercise 1: Check your understanding

1. **GS1** In which direction would you travel from Leongatha to reach Melbourne?
2. **GS1** In which direction would you travel from Coleraine to get to Melbourne?
3. **GS1** Provide three reasons why country *places* are important.
4. **GS4** Classify each of the characteristics in **FIGURE 1** as economic, social or *environmental*.
5. **GS5** Identify the characteristics in **FIGURE 1** that are attractive to you. Are most of these characteristics social, economic or *environmental*?

7.6 Exercise 2: Apply your understanding

1. **GS6** What are three opportunities in your community for young people to feel socially connected?
2. **GS5** Study the data in **TABLE 1**.
   (a) The percentage of the population under 20 is predicted to _____ in all municipalities.
   (b) Country regions are predicted to have a big increase in the percentage of population over 65. True or false? What is the evidence to support your choice?
   (c) Apart from the age of the population, suggest another difference in the demographic between country towns and big cities.
3. **GS6** In Australia, the trend is for people to move away from the country to the major cities. Suggest three reasons why you think this happens.
4. **GS1** Country towns are critical to the rest of Australia. Explain what role country towns can play on a national scale.
5. **GS6** How might governments use demographic information from country towns?

Try these questions in learnON for instant, corrective feedback. Go to www.jacplus.com.au.

7.7 Seasonal places

7.7.1 Falls Creek — a seasonal village

Many aspects of the way people live are influenced by the seasons. Our recreational activities, transport, housing features and clothing often change throughout the year. In some towns, there are changes in business activity and population as seasons change.

FIGURE 1 Location of Falls Creek, a ski resort in Victoria

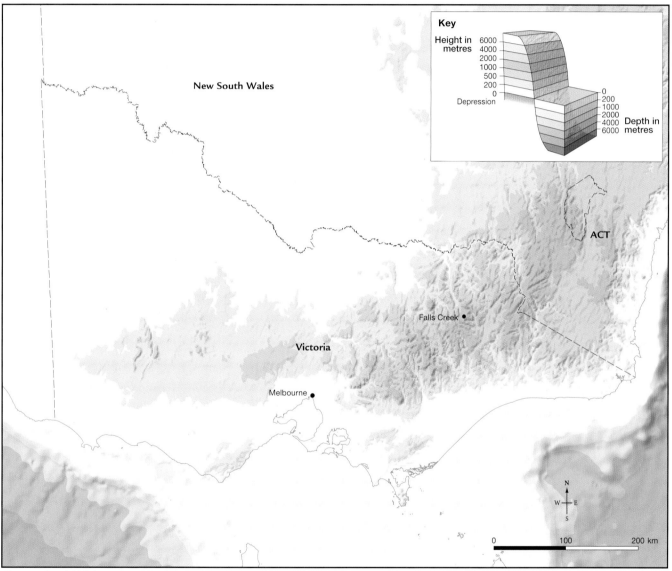

Source: Spatial Vision

Falls Creek is located in north-east Victoria about 20 kilometres south of Victoria's highest mountain, Mt Bogong. It is in the Great Dividing Range, which extends from northern Queensland to western Victoria. As a ski resort, Falls Creek is very busy in winter and quiet in the summer. The village sits at an **elevation** of 1765 metres above sea level.

7.7.2 Climate

Falls Creek is much cooler and wetter than places at a similar latitude. It holds several rainfall and temperature records in Victoria.

- Lowest temperature: −11 °C (3 July 1970)
- Highest monthly rainfall: 989.6 mm (July 1964)
- Highest annual rainfall: 3738.5 mm (1956)

The elevation of Falls Creek means the air is always cooler, which increases the chances of rain or snow. The southern **aspect** of the ski slopes means that the snow remains on the slopes for longer.

FIGURE 2 Elevation and temperature: the higher a place is above sea level, the less heat it receives.

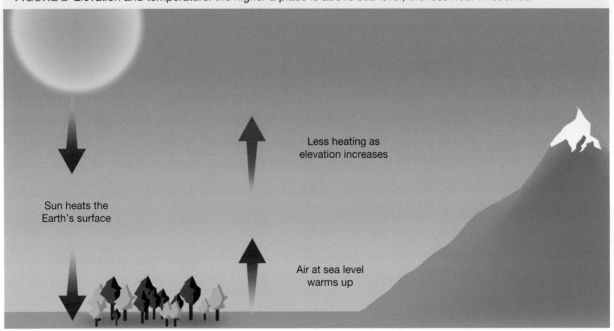

FIGURE 3 The influence of aspect on snow cover. The slope that does not face the sun is always cooler.

FIGURE 4 Climate graph for Falls Creek (latitude 36.87°S)

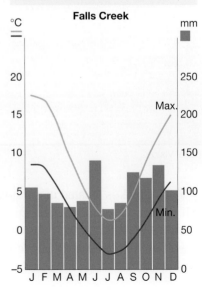

Snow is the major attraction at Falls Creek, and many snow-making machines are now used to improve snow cover and extend the ski season. Nearby Rocky Valley Lake provides plenty of water to feed into the high-pressure snow gun, but the temperature must also be very low (around −2 °C) for the snow machine to work.

7.7.3 Seasonal activities

FIGURE 5 Seasonal activities include skiing and bushwalking.

Winter

Clothing must protect the skin from wind, water and cold and the face from sunburn.

Summer

Clean air and mild temperatures suit outdoor activities such as bushwalking.

TABLE 1 Differences between summer and winter at Falls Creek

Data	Summer	Winter
Population*	225 workers plus visitors	2500 workers plus visitors
Visitor numbers#	35 000	130 000
Examples of type of work	Maintenance, building, road construction, revegetation and outdoor recreation	Maintenance, hospitality, snow activities, ski patrol and child minding
School enrolment*	10	40
Examples of activities	Cycling, bushwalking, fishing, altitude sports training, horse riding and tennis	Downhill skiing, cross-country skiing, snowboarding, snowshoeing, snow tubing, night skiing and snow bike riding
Special activities	Dragon boat races and alpine cycling competitions	Ice plunge, sled-dog races and skiing

Notes: *Numbers are approximate.
   #Many visitors stay for several days.

 Resources

 Interactivity Alpine regions — open or closed? (int-3093)

7.7 INQUIRY ACTIVITIES

1. In a group, brainstorm then create a list of towns that are busy in one season of the year. Choose one from the list and describe its location in at least three ways. In which season is the town busy? What is the attraction that draws people to the town? **Examining, analysing, interpreting**
2. Present the differences between summer and winter at Falls Creek in a visual format. Consider recreational activities, transport options, housing needs, clothing requirements and access to goods and services. **Comparing and contrasting**

7.7 EXERCISES

Geographical skills key: GS1 Remembering and understanding **GS2** Describing and explaining **GS3** Comparing and contrasting **GS4** Classifying, organising, constructing **GS5** Examining, analysing, interpreting **GS6** Evaluating, predicting, proposing

7.7 Exercise 1: Check your understanding

1. **GS1** In which state of Australia is Falls Creek?
2. **GS1** What is the special attraction that draws people to Falls Creek?
3. **GS1** According to **TABLE 1**, how many more people visit Falls Creek in winter than in summer?
4. **GS2** Using the description of the location of Falls Creek as a guide, describe the location of your town or city. Refer to your state, directions, distance and at least one other feature.
5. **GS5** Refer to the climate graph (**FIGURE 4**) for Falls Creek. In which months would you expect it to snow?
6. **GS2** Explain why July would be a suitable month to operate the snow-making machines. Which other months could also be suitable?

7.7 Exercise 2: Apply your understanding

1. **GS4** In which season is each of the following jobs most likely to be available at Falls Creek?
   (a) Child minder
   (b) Mechanic
   (c) Fitness trainer
   (d) Tennis coach
   (e) Ski instructor
   (f) Bicycle mechanic
   (g) Chef
   (h) Cleaner
   (i) Stable hand
   (j) Retail assistant
2. **GS6** How do the seasons influence the way of life in your town? Identify three aspects that *change* according to the season.
3. **GS6** Over the past 50 years, there has been a general trend of a slight decrease in snowfall. If this trend continues, what impact will it have on the village of Falls Creek? Consider the impact on people (social impact), on business (economic impact) and on the natural surroundings (*environmental* impact).
4. **GS6** Falls Creek has internet and mobile phone coverage. What influence do you think this would have on access to goods and services?
5. **GS6** What kind of new activities would you introduce to Falls Creek to boost tourism in the summer months?

Try these questions in learnON for instant, corrective feedback. Go to www.jacplus.com.au.

7.8 Places of change

7.8.1 On the move

A town will change over time if the factors influencing people's decision making about living there also change. Change may be due to government plans, the perception of the natural environment, the economic activities that are carried out in the place and access to resources and other places.

The original buildings in Tallangatta, in north-east Victoria, about 40 kilometres from Albury and Wodonga, can be seen only when the water level in Lake Hume is very low. The current town was moved from its original location in 1956. Houses were lifted onto trucks (with parts of the buildings often falling off during the journey) and moved about eight kilometres (see **FIGURE 1**). The original site, in a valley beside the Mitta Mitta River, was flooded when the size of Lake Hume was increased.

FIGURE 1 A Tallangatta house being moved to the new town site

7.8.2 Town closed

In 1917, it was decided that a town was needed on the dry and very warm Nullarbor Plain to provide services for the Indian Pacific railway (see **FIGURE 2**). With a population of 300, the town of Cook was once big enough to have a school, hospital, shop and accommodation for train drivers. When the railways were privatised in 1997, the town was closed. There are currently no known residents of Cook and it has effectively become a ghost town.

FIGURE 2 The location of Cook

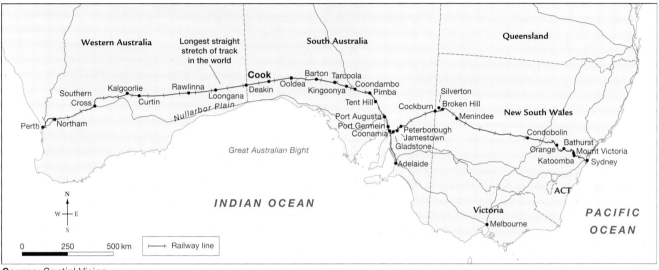

Source: Spatial Vision

7.8.3 Access to resources

Resource depletion

Silverton, 25 kilometres north-west of Broken Hill, was once home to 3000 people who mainly worked in mining (see **FIGURE 3**). Most people left, often taking their homes, when richer mines opened at Broken Hill. According to the 2016 census, the population of Silverton is 50, and the town is now visited by many tourists. The town and its semi-arid surroundings have been used as the setting in many films, such as *Mad Max 2, Dirty Deeds, Mission Impossible 2 and The Adventures of Priscilla, Queen of the Desert*. In the coming years, we may see another change in Silverton's population with the construction of the Silverton Wind Farm. After years of delays, construction of the renewable energy project is due to commence in 2019.

FIGURE 3 The hotel in Silverton and the surrounding landscape have been used in films such as *Mad Max 2*.

Resource discovery

Karratha is a hot, dry place 1600 kilometres north of Perth. It was founded in the 1960s for workers on the growing iron ore mines in the Pilbara region. In the 1980s the development of the natural gas industry encouraged further growth. The town currently supports about 22 000 people and is expected to support up to 40 000 by 2030.

DISCUSS

Discuss the following statement with a partner:
   '***Environmental*** factors are the main reason why towns ***change***'.
   Compose a clear paragraph to express your opinion. The first sentence will clearly state your view. The rest of the paragraph should contain at least two pieces of evidence to support your view.

7.8.4 Sea change

Margaret River, 270 kilometres south of Perth, has become popular because it offers a rural lifestyle and is accessible to the capital city. People who move from the city to the coast are said to have made a 'sea change'. Those who move to an inland location are said to have made a 'tree change'.

FIGURE 5 The Bussell Highway — the main street in Margaret River — 1991

FIGURE 4 The planned town of Karratha

Change over time

Many people now recognise that the Margaret River region has many attractions, such as beaches, waterways, caves, wineries, national parks and mild weather that suits farming and tourism.

However, what people have thought about the region has changed over time. Before 1830, the Noongar people, including the Wardandi Nation, valued the natural characteristics (such as flora, fauna, weather, sea and rivers) and made few changes to the natural environment. In 1830, white settlers arrived to cut down trees and sell timber. In 1950, they began using the cleared areas for dairy cattle and beef cattle.

Tourists also began to value and visit the region's natural features, such as beaches, rivers and caves. By 1970, people were moving from the city to enjoy the quiet country atmosphere and, by 1990, the area had become popular as a sea change destination.

TABLE 1 Population change in Margaret River

Year	Population of town
2001	3627
2006	4415
2011	5314
2016	7654
2021*	8500

*predicted population

TABLE 2 Origin of people who moved to Margaret River, 2006–2011

Previous place of residence	Number
New South Wales	83
Victoria	35
Queensland	44
South Australia	30
Western Australia	1004
Tasmania	13
Northern Territory	20
Overseas	365

7.8.5 Tourism

Port Douglas, 60 kilometres north of Cairns, was a busy port in the 1870s and had a population of more than 10 000. The mining that had attracted people to this hot, wet area did not last. By the 1960s, the population was only 100. In the 1980s, road and air access to the town improved. People were prepared to travel long distances from within Australia and from overseas to enjoy the warm weather, stunning beaches and the World Heritage areas of the Great Barrier Reef and Daintree rainforest. The permanent population is now about 3500. During the peak holiday season (May to November) the population of Port Douglas can increase to more than double its regular size.

FIGURE 6 Port Douglas in 1971, before the tourist boom

FIGURE 7 Port Douglas in 2019

DISCUSS

Many places change over time. Study the two photographs of Port Douglas in 1971 and in 2019.
  Discuss how the following people might respond to change that has taken place here:

- A resident whose family has lived in Port Douglas for three generations
- A shop owner
- A travel agent
- A tourism company owner
- A fisherman
- A painter/photographer.

[Personal and Social Capability]

7.8.6 Change in the future

Even in a small state like Victoria, predicted population growth varies across the state. Towns relying on big farms are predicted to lose population. The use of machinery and the closure of processing plants have reduced employment opportunities. Towns in regions very close to Melbourne are predicted to grow. People who live in these places still have access to jobs and entertainment in Melbourne even though they live in regional Victoria. More people means there is a need for more businesses.

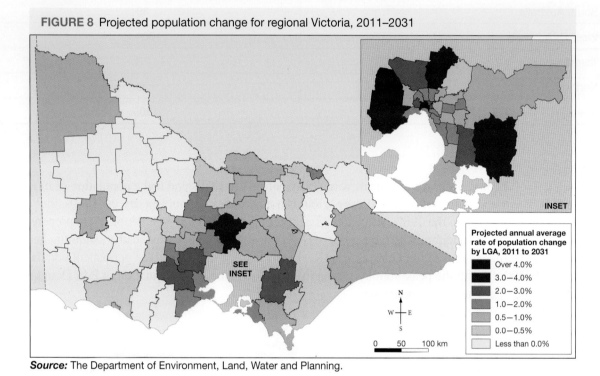

FIGURE 8 Projected population change for regional Victoria, 2011–2031

INSET

SEE INSET

Projected annual average rate of population change by LGA, 2011 to 2031
- Over 4.0%
- 3.0–4.0%
- 2.0–3.0%
- 1.0–2.0%
- 0.5–1.0%
- 0.0–0.5%
- Less than 0.0%

0   50   100 km

Source: The Department of Environment, Land, Water and Planning.

on Resources

Google Earth   Tallangatta
Silverton
Karratha
Margaret River

7.8 INQUIRY ACTIVITIES

1. Find maps of Victoria that provide information about landform and climate. Refer to your maps and **FIGURE 8** to complete the following.
   (a) Think about landform and population *change*. Are most areas of declining population in *places* that are not mountainous? Are most areas of increasing population on the coast side of the mountains?
   (b) Think about climate and population *change*. Are most of the highest growth population areas in *places* where rainfall is over 600 millimetres per year? Are most areas of declining population in *places* where the rainfall is lower?
   (c) What might be reasons for your findings in (a) and (b)?   **Examining, analysing, interpreting**
2. (a) Draw a sketch map of natural features at Port Douglas in 1971. Show the ocean, promontory, Dickson Inlet, flat land and hills. Add the settlement features (such as housing, roads and marina).
   (b) Using another colour or an overlay, show the settlement features for 2019.
   (c) Annotate your map to describe the *changes* that have occurred and the *changes* you think will happen in the next ten years.   **Classifying, organising, constructing**

3. Look at **FIGURE 5**. What do you think the main street in Margaret River looks like today? Predict what you would expect to see in terms of the road, footpaths, plants, types of shops, types of buildings and open spaces. Use Google Street View and check your hypothesis. Do you think Street View visited at a busy holiday time?

Evaluating, predicting, proposing

7.8 EXERCISES

Geographical skills key: GS1 Remembering and understanding **GS2** Describing and explaining **GS3** Comparing and contrasting **GS4** Classifying, organising, constructing **GS5** Examining, analysing, interpreting **GS6** Evaluating, predicting, proposing

7.8 Exercise 1: Check your understanding

1. **GS1** Which water storage drowned old Tallangatta? How far was the town moved?
2. **GS1** In which region of Western Australia is Karratha?
3. **GS1** What now draws people to Silverton?
4. **GS1** What is the population of Port Douglas in the peak holiday season?
5. **GS5** Study **FIGURE 2**.
   (a) How many states does the Indian Pacific railway travel through?
   (b) Why do you think the railway is called the Indian Pacific?
   (c) In which general direction does the train travel from Sydney to Perth?
6. **GS4** Read the description of the *change* over time for Margaret River. Create a timeline to show the *changing* view of Margaret River.

7.8 Exercise 2: Apply your understanding

1. **GS5** Refer to **TABLES 1** and **2** and calculate the population increase between each census.
   (a) In which time period was the greatest population increase in Margaret River?
   (b) What were the three main *places* new residents came from to settle in Margaret River between 2006 and 2011?
2. **GS6** Suggest why May to November is the peak holiday season in Port Douglas.
3. **GS6** Refer to the map in **FIGURE 8**.
   (a) Describe the location of the areas predicted to grow by more than 3 per cent. For, example, are they inland or by the coast? Are they in the north, south, east or west of the state? Are they clustered together or spread out? Are they close to Melbourne?
   (b) What will happen to towns in regional Victoria?
   (c) Estimate the proportion of Victoria that is predicted to increase its population and the proportion that is predicted to decrease its population.
4. **GS3** Factors that cause *change* can be categorised as social (related to people), economic (related to money) or *environmental* (related to setting or surroundings). Consider all the reasons for *change* provided in this subtopic and list each in its correct category.
5. **GS6** What would be the advantages and challenges of living in a town such as Port Douglas, which relies on tourism? Use speech bubbles like the ones below.

It would be good because …

But …

Try these questions in learnON for instant, corrective feedback. Go to www.jacplus.com.au.

7.9 Isolated settlements

7.9.1 Way up north

Life in some towns is strongly influenced by geography. It is now possible to live in extreme conditions and be socially connected while enjoying access to a wide range of goods, services and community activities.

Dawson, with a population of 1400, is the second largest city in the Canadian state of Yukon. At the latitude of 64°N, it is only about 360 kilometres from the Arctic Circle.

Dawson is a long way from neighbouring towns. It is 770 kilometres from the next town to its north (Inuvik) and 810 kilometres to Anchorage, the largest town in Alaska.

FIGURE 1 Location of Dawson, Yukon, Canada

Source: Spatial Vision

7.9.2 Climate

TABLE 1 Average hours of sunlight at Dawson, Yukon and Longreach, Queensland

Month	Dawson, latitude 64°N	Longreach, latitude 23°S
January	4	13
February	6.9	13
March	10.2	12.5
April	13.7	12
May	17.1	11
June	20.6	11
July	21.4	11
August	18.1	11
September	14.6	12
October	11.2	12.5
November	7.8	13
December	4.7	13.5

7.9.3 Settlement

The site of Dawson, on the junction of the Yukon and Klondike rivers, was always an important hunting and meeting site for indigenous peoples, who still live in the area. The harsh climate deterred white colonisers from the area until gold was discovered in 1896. The population grew by thousands every week, quickly reaching 30 000.

However, most people did not stay long. Living conditions were very poor: people lived in tents and huts; they had no power, water or sewerage; and crops would not grow in the low temperatures. The **permafrost** also made it difficult to dig the foundations for buildings. Furthermore, it turned out that the gold was not accessible: it was in gravel that was frozen for most of the year.

Today

Today the town has **infrastructure**, and housing is solid and heated. The community of Dawson operates year round, with the climate having a strong influence on activities. A road to Alaska has encouraged tourism, and new mining techniques have increased goldmining.

The weather conditions mean that some jobs, such as road-making and building, cannot be done in the winter. Some of those employed in these industries move south in the winter, either for work or holidays.

The town of Dawson is no longer completely isolated in winter. Internet and phone connections allow communication; snow-ploughing and spreading of sand keep the roads open most of the time; and planes fly in and out all year.

The weather influences many aspects of social life in Dawson. The 'Thaw di Gras' carnival celebrates the end of winter, and music festivals and a kayak marathon take place in summer. Winter is the time of snowmobile treks, skiing and the famous Yukon Quest dog-sled race from Whitehorse (Yukon) to Fairbanks (Alaska). During the race, dogs and mushers (sled drivers) need protection from the cold, but if the weather warms to −4 °C, it becomes too hot for the dogs to travel during the day.

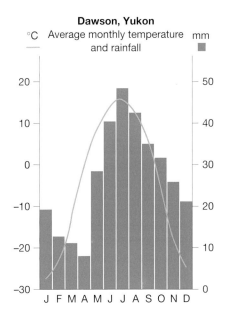

FIGURE 2 Climate graph for Dawson

FIGURE 3 It is so cold in winter that the Yukon River freezes over, and the ice is thick enough to support fully loaded trucks.

7.9 INQUIRY ACTIVITIES

1. (a) Construct a multiple bar or column graph to show the average number of hours of sunlight at Dawson and Longreach.
   (b) On your graph, use colour, shading or symbols to show summer and winter at each location.
   (c) Write one statement that is true for both Dawson and Longreach about their hours of sunlight.
   (d) Imagine you live in Dawson, where the school year is from August to June. In which months would you arrive or leave in the dark? **Classifying, organising, constructing**
2. Use the **Bureau of Meteorology** weblink in the Resources tab to find the average monthly temperatures for your location. Present the statistics for the minimum and maximum temperature in a graph. Compare your graph to the one provided for Dawson. Write three clear statements comparing the average monthly temperatures of Dawson and your location. Write three clear statements comparing the monthly rainfall for Dawson and your location. **Classifying, organising, constructing**

7.9 EXERCISES

Geographical skills key: GS1 Remembering and understanding **GS2** Describing and explaining **GS3** Comparing and contrasting **GS4** Classifying, organising, constructing **GS5** Examining, analysing, interpreting **GS6** Evaluating, predicting, proposing

7.9 Exercise 1: Check your understanding

1. **GS5** Refer to the map in **FIGURE 1**. How far is Dawson from Whitehorse, the state capital?
2. **GS1**
   (a) What influenced indigenous people to live in the Dawson area?
   (b) Why did people move to Dawson in 1897?
   (c) What influences people to live in Dawson nowadays?
3. **GS2** List at least four ways in which the natural **environment** has influenced the settlement of this area.
4. **GS2** What factors have improved liveability in Dawson?
5. **GS2** Define the term 'infrastructure' and list three examples.

7.9 Exercise 2: Apply your understanding

1. **GS4** Activities in Dawson include volleyball, curling, film festivals, craft, dances, motorcycle riding, gold panning, softball, writing workshops and outhouse races. Which of these would occur only in summer? Which would occur only in winter? Which could occur all year?
2. **GS5** Which months of the year would be most popular for tourists visiting Dawson City? Provide at least two reasons to support your answer.
3. **GS3** How do you think the climate would make school life in Dawson different from your life?
4. **GS6** What do you think happened to the early buildings when the heating from houses warmed the soil and melted the permafrost?
5. **GS5** How have people tried to overcome the disadvantages of this location?

Try these questions in learnON for instant, corrective feedback. Go to www.jacplus.com.au.

7.10 A nomadic way of life

7.10.1 Traditional Tuareg way of life

Most people live in the one place, but from time to time they may move to a new location. There are about 30 million people in the world who live a nomadic lifestyle. Nomads do not wander aimlessly. From time to time they pack up all their possessions and move, often returning to a place at some point in the future.

The Tuareg people (pronounced twah-reg) lead a nomadic way of life mainly in the Sahara Desert. A number of related families live in groups of 30 to 100. These groups usually move to a new site every two or three weeks, because the environment provides very little water and food.

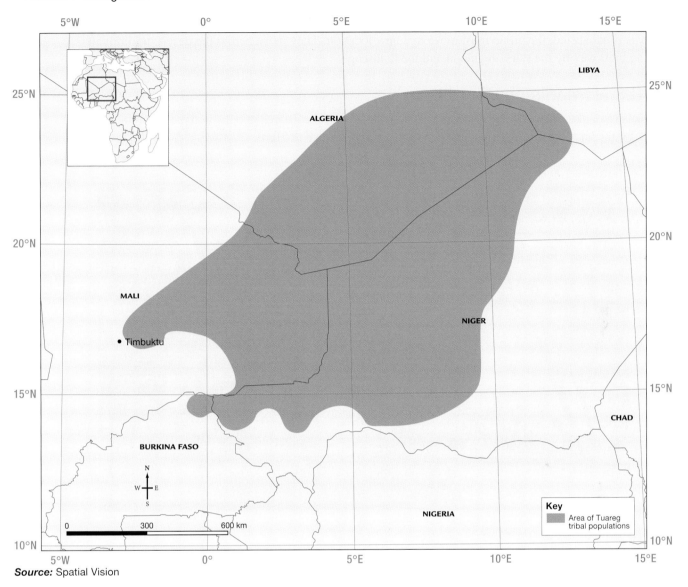

FIGURE 1 Tuareg areas

Source: Spatial Vision

7.10.2 A harsh environment

Timbuktu is a town in Mali (see **FIGURE 1**) that has long been famous as a trading town. Here, goods such as salt were brought from the north across the Sahara, and goods such as gold came from the south.

Daytime temperatures are high and night-time temperatures can be cold. Rainfall is very low and unreliable. There can be long periods without rain and there can be sudden heavy downpours. Strong winds sometimes cause sandstorms that turn the sky yellow or orange, reduce visibility and cover everything with sand.

TABLE 1 Average temperature and rainfall, Timbuktu

	Jan.	Feb.	Mar.	Apr.	May	Jun.	Jul.	Aug.	Sep.	Oct.	Nov.	Dec.
Average max. temp. (°C)	31	35	38	41	43	42	38	35	38	40	37	31
Average min. temp. (°C)	13	16	18	22	26	27	25	24	24	23	18	14
Average rainfall (mm)	0.5	0.5	0.5	1	4	20	54	93	31	3	0.5	0.5

Source: http://www.timbuktu.climatemps.com/temperatures.php

The climate means there is little vegetation and very little water in Timbuktu. Significant plant growth is only found at oases. This means that families have difficulty finding wood for cooking fires or vegetation for their animals. When firewood is unavailable, dried camel dung is used as fuel. When food for the animals is exhausted, the group must move to a new location. They move across the desert, traditionally finding their way by the stars, the moon and the landscape.

7.10.3 Nomadic herding

Animals have always been the most important possessions of the Tuareg people, and the need for grass and water for the animals is why the Tuareg people move from place to place.

FIGURE 2 Tuareg camels and goats

7.10.4 Household possessions

A nomadic way of life means that Tuareg families do not have many possessions, and the few they have must be light and portable.

For example, their housing is usually in the form of tents that have a simple structure. These provide shelter from heat in the daytime and from cold at night-time. Tuareg tents used to be made of skin and woven cloth but now they are often made of nylon. They are always placed so that the doorway is facing the non-wind side.

The main furnishings in their tents are rugs to cover the ground and to provide a sleeping mat. Sometimes a bed is constructed from palm slats resting on thin logs. Other furniture includes cushions and drawers or chests. A tent and other belongings can be packed up within a few hours.

FIGURE 3 A Tuareg woman making tea outside her family tent in Niger

For cooking, the Tuareg people use an open fire, sometimes with a hotplate. They use only a few utensils such as pots, containers, plates and spoons. Eating utensils are also minimal, and sometimes food is eaten by hand from a communal plate, as is customary across many different cultures around the world.

Tea is the most popular drink, and is the only food that includes sugar. A family will usually have a teapot and small cups, and the making of tea is a ceremony. Tea is offered to visitors and becomes part of business discussions.

7.10.5 Clothing

The Tuareg people always wear clothing that covers them from head to toe. The men, who spend more time outside, wear a headscarf to cover their head, neck and much of their face. This is to prevent sunburn, stop the lips from cracking, and slow down the drying out of the mouth. Men over 18 years of age traditionally wear blue headscarves. Their long clothes also provide warmth during the coolest time just before sunrise.

7.10.6 Changes to way of life

In recent times, the Tuareg people have been forced to change their way of life for a number of reasons:
- Drought has reduced the amount of food available for the animals.
- Private ownership of land has reduced the areas in which the Tuareg can move.
- Political unrest has made some areas unsafe.
- Population growth has placed pressure on the available land.

An increasing number of Tuareg people are moving to the south and becoming semi-nomadic. This means they are in one location for a large part of the year. They are building more permanent buildings, such as **adobe** houses, and using some irrigation to help crops grow. Working for money is becoming more common, and children are sometimes able to attend school. Healthcare, such as the provision of vitamins to improve nutrition, is now reaching more people. Solar panels are being used in some areas to produce power to charge mobile phones, run solar cooking ovens and provide lights in schools.

Some traditions remain. Nomadic herding is still valued as the most important activity, and tents are still the main form of housing. Loose clothing is still popular, and the men are still well known for their blue scarves.

on Resources

✦ **Interactivity** Tuareg treks (int-3094)

7.10 INQUIRY ACTIVITIES

1. Refer to an atlas and complete the following sentence about where the Tuareg people live.
   (a) For most of the region where the Tuareg people live, the population per square kilometre is _____.
   (b) Annual rainfall is _____.
   (c) Agriculture is _____.
   (d) GDP is _____.
   (e) Average January temperatures are _____.
   (f) Average July temperatures are _____.   **Examining, analysing, interpreting**
2. Select another group of people who lead a nomadic way of life (such as the Awa, Penan, Orang Rimba or Tibetan nomads). Your task is to gather information about your chosen group. To guide your research, develop a key question for each of the following criteria: the environment, the people and the lifestyle. Represent your findings in at least three formats (such as a graph, a map, a satellite image, a photo, text or a diagram). Write three clear sentences to compare your group to the Tuareg people.
                                                          Comparing and contrasting

Geographical skills key: GS1 Remembering and understanding **GS2** Describing and explaining **GS3** Comparing and contrasting **GS4** Classifying, organising, constructing **GS5** Examining, analysing, interpreting **GS6** Evaluating, predicting, proposing

7.10 Exercise 1: Check your understanding

1. **GS5** Refer to **FIGURE 1**.
   (a) Name the five main countries through which the nomadic Tuareg people move.
   (b) On which continent do the Tuareg people live?
   (c) Do the Tuareg people live in the northern hemisphere or the southern hemisphere?
2. **GS1** Where is Timbuktu located?
3. **GS5** Refer to **TABLE 1**.
   (a) What is the average yearly total rainfall in Timbuktu?
   (b) What is the average maximum temperature in the coolest month?
4. **GS2** Identify and explain three ways in which the environment has influenced the traditional way of life of the nomadic Tuareg people.
5. **GS2** Provide four reasons to explain why animals are highly valued by the Tuareg people. Suggest an item that is highly valued in Australian culture.

7.10 Exercise 2: Apply your understanding

1. **GS6** The Tuareg people may be required to change their way of life in the future. What do you think will happen to the Tuareg's nomadic way of life? Consider the influence of **environmental** factors (related to the natural world), economic factors (related to businesses and work) and social factors (related to people's welfare, hopes and attitudes).
2. **GS3** Compare a day in the life of a Tuareg nomad and yourself. Include:
   (a) where and how you both live
   (b) diet, clothing, housing type, possessions, settlement size, schooling, and travel.
3. **GS2** What are adobe houses?
4. **GS2** An increasing number of Tuareg people are becoming semi-nomadic. What does this mean?

Try these questions in learnON for instant, corrective feedback. Go to www.jacplus.com.au.

7.11 India, past and present

7.11.1 What is Old Delhi like?

Old Delhi is an area within the modern city of New Delhi in India. Old Delhi consists of the original walled city that dates back to 1639. It was founded by the Mughal emperor Shahjahan, and was known then as Shahjahanabad. The local people pride themselves on being a peaceful community in which Muslims, Hindus and Christians have lived together side by side for hundreds of years.

The British began developing the area now known as New Delhi outside the city walls of Old Delhi in about 1911. However, life has continued within the walls of the 6.1-square-kilometre old city, which still has its original 14 gates. This makes it just a little smaller than Lake Burley Griffin in Canberra, but with a population density of over 25 000 people per square kilometre!

FIGURE 1 Building a new multi-storey building inside Old Delhi

FIGURE 2 Even though mobile phones are found everywhere in India, there is still room for a business that offers you a telephone line when you need it. The high buildings in Old Delhi mean that mobile reception is not good in many locations within the city walls.

FIGURE 3 People of different religions have lived in Delhi for centuries. About 80 per cent of the inhabitants of Old Delhi are Muslims, whereas in the whole of Delhi, about 80 per cent of the population are Hindu. A number of **mosques**, temples and Christian churches are crammed into the old city, including (a) the Jama Masjid (Delhi's largest mosque); (b) small but important mosques such as Kalan Masjid, built around 1387; (c) the Holy Trinity Church.

FIGURE 4 The bazaars in Old Delhi sell everything from food to bicycles.

FIGURE 5 Old Delhi is a hub of small industries, from metalwork to craft shops and food preparation. Power and telephone lines are draped from building to building.

FIGURE 6 As with many other small shops in Old Delhi, the welder (a) works with old homemade equipment. (b) The cobbler repairs shoes with a minimum of equipment, using his feet as a vice. (c) The tailor uses an old sewing machine, operated by a foot pedal.

FIGURE 7 The narrow streets make rubbish collection impossible, even if local government did offer it. The roofs of lower buildings often become rubbish dumps for those on the higher floors. In Delhi, nearly all rubbish collection is handled by private contractors or by the wastepickers.

FIGURE 8 Getting ready for iftar, the communal **Ramadan** festival after sunset. The place is full of a range of chaat (tangy and spicy snacks). During Ramadan, communities start to cook a fastbreaking meal in the afternoon so that it is ready for all to eat after sunset. This often takes the form of a neighbourhood banquet, celebrating friendships, family and community.

FIGURE 9 Access to clean water is a major problem in Old Delhi. There is no internal plumbing, although some households are able to obtain water through a courtyard distributor system. In general, water supply is communal in Old Delhi, just as it is in much of rural India and older urban areas. Usually, only the middle class and wealthy can afford to have running water. The courtyards of newer homes sometimes have a water tank on the roof.

FIGURE 10 Schools in Old Delhi exist in small buildings down side alleys. Some schools teach in the English language.

FIGURE 11 With few jobs around, young people learn to be creative in finding a way of making a living. They work in the **informal sector**. Young boys might sell cool drinks, with cordial made with water from an urn. This is then cooled by ice from a block, which is big enough to take all day to melt.

7.11.2 How is modern India changing?

The multiple realities of India

Modern India is a very complex place. Indians themselves often talk about the 'multiple realities' of India. This means that there are many different pictures that visitors might see of life in India. Some of these may contradict each other, yet all of them are true.

There is a great difference between modern India and traditional India. Modern India is in the top ten national economies of the world. Since the 1990s, government policies have encouraged industrialisation, which has given jobs to many people and money for them to spend. India has over 1.3 billion people, and there is now an increasing number of middle-class citizens who are highly educated, earning high incomes, and wanting the type of lifestyle that money can bring. In 2018, India had 104 billionaires (in US dollars), and almost 5 per cent of the world's billionaires.

However, the gap between rich and poor is increasing. Economic growth has decreased the number of people living in poverty, but over 40 per cent of the population still lives in very basic conditions. Around 40 per cent of people still work in agriculture, and the rural poor are substantially worse off than other Indians. Incomes for farm workers are well below average.

A large proportion of Modern India's economic growth has been in the service sector, where people work in jobs providing skills ranging from gardening and laundry to accountancy. For instance, many businesses in North America, Europe and Australia have 'outsourced' their telecommunications and work to call centres in India. This means that they hire people in India to do jobs that are more expensive to have done locally. Many global companies now have bases in India.

FIGURE 12 In rural areas, many of the poor work in the informal sector. This snake charmer is one of a dying profession, working mainly for tourists and in festivals and markets, despite the fact that it is now illegal. The serpent is sacred in Hinduism, and snake charmers were once thought to have gifts of healing.

FIGURE 13 A long drought in the 2000s put great pressure on dwindling water supplies in India, particularly in the countryside. However, urban growth and few environmental controls on local industry have also led to the pollution of small yet important lakes, like this one near Agra.

FIGURE 14 The contrast of modern India: (a) When modern buildings are constructed, bamboo poles are still used as scaffolding. (b)–(d) The growing middle class has created a 'westernised' lifestyle for many Indians, where people place great importance on always buying new goods and services. This can be seen in the advertisements within large shopping centres. (e) There is still room for local customs, however: look at what is on the sign hanging above the front of the McDonald's store. Is this type of customer service what you would find at your local McDonald's store?

FIGURE 15 Three methods of transport in modern India

on Resources

Weblinks   Wordle

               Delhi passport

               Indian Premier League

 Google Earth Old Delhi

7.11 INQUIRY ACTIVITIES

1. Using the SHEEPT system of classification, construct a table like the one below to summarise the major features of Old Delhi and of the lives of the people who live there. **Classifying, organising, constructing**

	Social features	Historical features	Economic features	Environmental features	Political features	Technological features
Feature 1						

2. Using the table that you have drawn up, identify five characteristics of Old Delhi that are very different from the place in which you live, and five characteristics that are similar. **Comparing and contrasting**
3. Use Google Earth to view an image of the space of Old Delhi in the present day. Using the Google Earth options, create an image of Old Delhi showing the distribution of the major human features, such as churches and mosques. Print out a copy of the map you have created. **Classifying, organising, constructing**
4. One of the areas where the growth of the wealthy class in India can be seen is in sport. Use the **Indian Premier League** weblink in the Resources tab to research the structure of the Indian Premier League (IPL) in cricket. Create a map showing where the different teams are based in India. Construct a table of the names of the owners and how they created their fortunes. What conclusions can you draw about the IPL as a symbol of modern India? **Classifying, organising, constructing**

7.11 EXERCISES

Geographical skills key: GS1 Remembering and understanding **GS2** Describing and explaining **GS3** Comparing and contrasting **GS4** Classifying, organising, constructing **GS5** Examining, analysing, interpreting **GS6** Evaluating, predicting, proposing

7.11 Exercise 1: Check your understanding

1. **GS1** Old Delhi consists of the original walled city that dates back to which year?
2. **GS1** Which three religions have lived peacefully together for hundreds of years in Old Delhi?
3. **GS1** Old Delhi has an area of _____ square kilometres and a population density of over _____ people per square kilometre.
4. **GS1** What does the phrase 'the multiple realities of India' mean?
5. **GS1** What is the population of India?
6. **GS1** Complete the following sentence:
   The number of people in poverty is decreasing, but over _____ per cent of the population still lives in very basic conditions, and _____ per cent still work in agriculture.

7.11 Exercise 2: Apply your understanding

1. **GS4** Classify each of the following features of Old Delhi as social, historical, economic, environmental, political or technological.
   (a) More people, rather than machines, doing jobs.
   (b) Emphasis on wasting nothing in production.
   (c) Mosques and churches are important.
   (d) No motorised transport is used, except for the occasional motorbike.
   (e) Most of the buildings are old.
2. **GS6** Identify the main feature in the environment of Old Delhi that you believe to be the major problem facing its future. Justify your decision, using information from this subtopic and further information from internet searches and library resources.
3. **GS3** Compare the living conditions of Old Delhi with those that can be seen in the places illustrated in other subtopics. What similarities and differences do you notice?
4. **GS3** Compare life in Old Delhi with life in India today.
5. **GS6** What do you think is the most significant issue facing New Delhi today? Describe what impacts this issue has on people and what could be done to solve the problem.

Try these questions in learnON for instant, corrective feedback. Go to www.jacplus.com.au.

7.12 Thinking Big research project: Investigating change over time

SCENARIO

Have you ever been outside and looked at the landscape in front of you and wondered what this area looked like before humans existed? Have you ever wondered what once stood in the place where your house now stands, or your school?

As we've learnt in this topic, it is possible for places to change over time. These changes can be caused by natural phenomena or they can be caused by human activity. You may have even seen some changes in your neighbourhood in your own lifetime. In this task, you will investigate how your local area has changed over time.

Select your learnON format to access:

- the full project scenario
- details of the project task
- resources to guide your project work
- an assessment rubric.

 Resources

 projectsPLUS Thinking Big research project: Investigating change over time (pro-0238)

7.13 Review

7.13.1 Key knowledge summary
Use this dot point summary to review the content covered in this topic.

7.13.2 Reflection
Reflect on your learning using the activities and resources provided.

 Resources

 eWorkbook   Reflection (doc-32141)

Crossword (doc-32142)

 Interactivity   A world of people and places crossword (int-7704)

KEY TERMS

adobe bricks made from sand, clay, water and straw and dried by the sun

aerial photograph a photograph taken of the ground from an aeroplane or satellite

aspect the direction in which something is facing

country the place where an Indigenous Australian comes from and where their ancestors lived; it includes the living environment and the landscape.

demographic describes statistical characteristics of a population

elevation height of a place above sea level

extensive land use land use in which farms are huge, with few workers and not many cows or sheep per hectare

incentive something that encourages a person to do something

informal sector jobs that are not officially recognised by the government as official occupations and that are not counted in government statistics

infrastructure the basic physical and organisational structures and facilities that help a community run, including roads, schools, sewage and phone lines

intensive land use land use in which farms are smaller but have more workers and machinery to produce high yields per hectare; examples are dairy and poultry farms, orchards, vegetables and feedlots.

mosque place of worship for people who follow Islam (Muslims)

natural resources resources (such as landforms, minerals and vegetation) that are provided by nature rather than people

permafrost permanently frozen ground not far below the surface of the soil

pull factors positive aspects of a place; reasons that attract people to come and live in a place

push factors reasons that encourage people to leave a place and go somewhere else

Ramadan the month of fasting in the Islamic calendar. It is a time for abstaining from food, drink and other physical needs during daylight hours, as well as reconnecting spiritually with God.

8 Liveable places

8.1 Overview

What does a place need to have to make it liveable? Is everyone's idea of liveability the same?

8.1.1 Introduction

Your quality of life is influenced by many factors, such as climate, landscape, community facilities, the location of your home, the sense of community identity and links to other settlements. You probably have an idea of a street, town, city or suburb where you would like to live, and your opinion may be quite different from those of others. This is because other people see different factors as important. This chapter will look at how people define and improve liveability.

on Resources

▢ **eWorkbook**   Customisable worksheets for this topic

▦ **Video eLesson**   Making places liveable for young people (eles-1621)

✎ **Google Earth**   Kolkata

LEARNING SEQUENCE
8.1 Overview
8.2 Defining liveability
8.3 Where are the most liveable cities?
8.4 **SkillBuilder:** Drawing a climate graph   `online only`
8.5 Melbourne — a liveable city
8.6 Liveability and sustainable living
8.7 Less liveable cities
8.8 Improving liveability
8.9 Liveable communities and me
8.10 **SkillBuilder:** Creating and analysing overlay maps   `online only`
8.11 **Thinking Big research project:** Liveable cities investigation and oral presentation   `online only`
8.12 **Review**   `online only`

To access a pre-test and starter questions, and receive immediate, **corrective feedback** and **sample responses** to every question, select your learnON format at www.jacplus.com.au.

8.2 Defining liveability

8.2.1 What do people think about liveability?

If you were told that Vancouver or Melbourne was the world's best place to live in, or the world's most liveable city, what would you think this means? Do city councils just brag about how good their city is or can liveability be measured? Is liveability the quality of life experienced by a city's residents?

Here are some examples of what fictional people think about the liveability of their community. They come from different places and they are all trying to explain what liveability means to them.

'I think a liveable city is a city where I can have a healthy life and where I can safely and quickly get around on foot or by bicycle, public transport or even by car — as a last resort. A liveable city is a city for everyone, including children and old people, rich and poor, and people of different religions, races and fitness levels. A liveable city should be attractive, and have good schools, a choice of things to do and fresh air.'

John from Perth

'I think that a place is liveable if I have food every day, I do not have to walk more than ten minutes to collect water for cooking and my father has work close by, so he is home for dinner. Liveability means warm weather, enough rain and being able to go to school every day.'

Nafula from Kenya

'Liveability is all about the **natural environment**. I think a place is liveable if the air is clean, there is plenty of water in the river and there is a healthy forest nearby. Being able to grow your own food, use renewable energy and live a simple life are all a part of what is important to me and can make a place liveable.'

Joy from Huon Valley, Tasmania

'A liveable place is somewhere I can have a computer and a television and a bed of my own in my own room. I would like a bike to get to school, three meals a day and two sisters. A liveable place would be clean, safe and modern. My grandmother and aunty would also live with us.'

Jing from a village in rural China

'A liveable community offers many activities, celebrations and festivals that bring all of its residents together. Every year at Carnevale, my whole neighbourhood comes together to dance the samba. I would never wish to live anywhere else.'

Raul from Rio de Janeiro

'Liveable cities have housing that is close to jobs, services and transport and is available for all income levels. Neighbourhoods are pedestrian-friendly with green spaces and lively retail sectors. They are mostly car-free, and have good schools and public buildings. A liveable city needs lots of different choices — choices in ways to live, places to work, shop and eat, and locations to linger in — whether alone or with other people.'

Alex, property developer from New York

'The place that I think would be the most liveable is Darwin. It has great footy grounds, public transport, good food, good houses, good shops and good schools. Where I live, my house is a dump and I cannot get anywhere unless I walk. I would like to live in Darwin and play football.'

Sam from near Alice Springs

'Liveability means that I have a good job, good food, a nice house, a newish car, nice neighbours and a community that cares about my family and me.'

Oscar from western suburbs of Sydney

'The community is what makes a place liveable. Being connected with my neighbours, through the community gardens, food co-op, volunteer network at our kids' school and the car-share scheme all make me feel a valued member of my community. I like knowing people who care and that we all care for each other.'

Laura from Bristol, United Kingdom

8.2 INQUIRY ACTIVITY

1. (a) Ask a much older person to describe the living conditions in the community they lived in as a teenager. Record or write down their memories.
   (b) Ask this older person how they would have measured liveability when they were young.
   (c) Do you think the current liveability of your community is better than that described by the older person? Provide examples to support your view. **Evaluating, predicting, proposing**

8.2 EXERCISES

Geographical skills key: GS1 Remembering and understanding **GS2** Describing and explaining **GS3** Comparing and contrasting **GS4** Classifying, organising, constructing **GS5** Examining, analysing, interpreting **GS6** Evaluating, predicting, proposing

8.2 Exercise 1: Check your understanding

1. **GS1** In your own words, define the term 'community'.
2. **GS1** What is liveability?
3. **GS2** Write a statement, similar to those in this subtopic, about the community that you live in and what makes it liveable.
4. **GS6** Would the statement of liveability in your community be different if you were blind, unemployed, elderly or unable to speak English? Write what you think a community liveability statement would be for two such residents of your community.
5. **GS5** Carefully read the different opinions about what makes a *place* liveable.
   (a) Make a list of the common themes mentioned by these people.
   (b) Is there a shared common definition of what makes a *place* liveable? Discuss.
   (c) On a map of the world, locate each *place* mentioned in this subtopic.
   (d) Does the *place* in which each person lives appear to influence their definition of the term *liveability*? Discuss.

8.2 Exercise 2: Apply your understanding

1. **GS6**

   (a) Think about your community 50 years from now. How will the characteristics of this **place** be different? For example, think about the type of houses, the **distribution** of houses, the amount and type of traffic, the age of the population, the community facilities and other characteristics you think will be significant.

   (b) What type of inventions might improve liveability?

2. **GS6** We have so far described liveability in a general sense. Sometimes living conditions can change quite quickly. Provide examples of how natural events, political events or economic events can influence living conditions.

3. **GS6** Which viewpoint or viewpoints about liveability do you agree with more than others? Give reasons for your answer.

4. **GS3** Examine each of the liveability statements and compile a list of factors mentioned. Which factors do you consider to be the top five? Justify your response.

5. **GS3** Identify three factors that you think are *not* important in determining liveability. Justify your choices.

Try these questions in learnON for instant, corrective feedback. Go to www.jacplus.com.au.

8.3 Where are the most liveable cities?

8.3.1 What is liveability?

Everyone likes to be able to tell you they are the best, or in the top ten of some category. Cities are no different. If you look at the official websites for many international cities, they will tell you that they are the safest, wealthiest, fastest-growing or have the best events calendar. Being able to boast that a city is the world's most liveable is great publicity.

Liveability can be defined as 'the features that create a place that people want to live in and are happy to live in'. It is usually measured by factors such as safety, health, comfort, community facilities and freedom.

8.3.2 Who says which is the most liveable?

Several international organisations have created lists of the world's most liveable cities. These organisations each compare data and produce a table that ranks the liveability of cities. This information is collected for workers considering overseas transfers or for companies that may need to compensate workers who are transferred to a low-ranked city. The figures can also be used to attract migrants or investment. The various rankings compare a large number of cities; however, not all cities in the world are included in each survey.

The criteria used to produce the rankings include:

- stability or personal safety (crime, terror threats and civil unrest)
- healthcare
- culture and environment (religious tolerance, corruption, climate and potential natural disasters)
- education
- infrastructure (transport, housing, energy, water and communication)
- economic stability
- recreational and sporting facilities
- availability of consumer goods (food, cars and household items).

FIGURE 1 shows the top ten and bottom ten in the global cities liveability rankings, as released by the Economist Intelligence Unit (EIU) in 2018. These rankings are released each year, so it is possible for you to log on (use the **Economist Intelligence Unit** weblink in the Resources tab) to get the most recent update to the rankings. This survey ranks 140 cities; a score of 100 equates to the perfect or ideal city. In previous years Vienna, Melbourne and Vancouver have shared the top ranking as the world's most liveable city. In 2018 Vienna took out the number one ranking with Melbourne pushed to second. Vancouver's ranking has fallen to sixth.

The map shows that many of the world's top cities have scores that are very similar. The difference in score between the top ten cities is only 2.5 points.

Between 2008 and 2018, the average global liveability score has increased by almost one percentage point. Of the 140 cities included in the liveability survey, half have improved their overall status. Competition at the top of the list is fierce. Four cities have been forced out of the top ten between 2017 and 2018.

FIGURE 1 The top ten and bottom ten in the global cities liveability rankings, as released by the Economist Intelligence Unit (EIU) in 2018

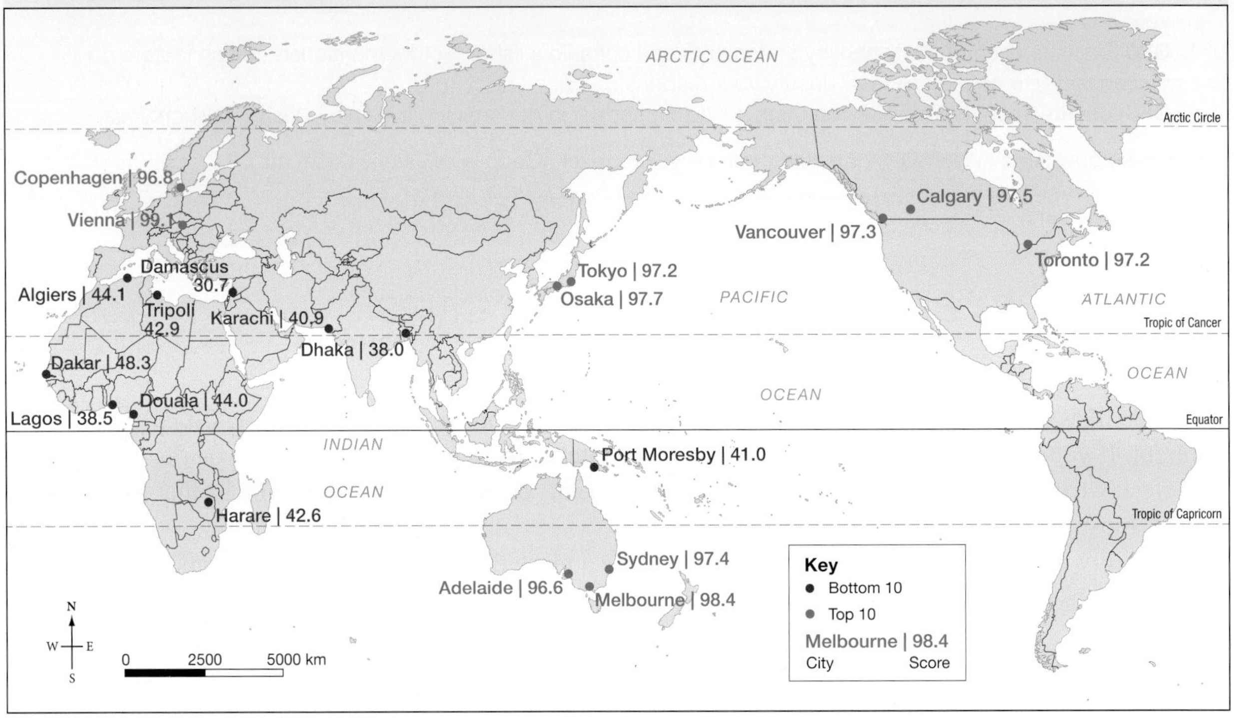

Source: Economist Intelligence Unit (EIU) 2018

TABLE 1 Changes in city liveability ratings from 2017 to 2018

City	2017 ranking	2018 ranking
Perth	7	14
Auckland	9	16
Helsinki	9	16
Hamburg	10	18

What do these top ten liveable cities have in common?

Looking at the locations of the most liveable cities, you can see most are found in Australia and Canada with three each; followed by Europe and Japan with two apiece. They are all mid-sized cities, have quite low **population density**, low crime rates and **infrastructure** that copes quite well with the needs of the local community. They are found in places where there is a **temperate climate**, perhaps with the exception of Toronto and Calgary, which do have very cold winters.

The top cities also tend to be modern cities, not much more than 300 years old. They have been planned so that people can travel around them by both public and private transport. They are also found in some of the world's wealthiest or most developed nations.

Australian and Canadian cities perform better than cities in the United States due to US cities' higher crime and congestion rates. The highest ranked United States city is Honolulu at 23.

 Resources

 Interactivity My most liveable country (int-3095)

 Weblink   Economist Intelligence Unit

Explore more with my**World** Atlas

Deepen your understanding of this topic with related case studies and questions.
● Investigating Australian Curriculum topics > Year 7: Place and liveability > **Polluted cities**

8.3 INQUIRY ACTIVITIES

1. Work with a partner or in a group to find the most recent population figures for each of the cities shown on the map in **FIGURE 1**. List your findings. Write one sentence to describe the population of the most liveable cities. Write one sentence to describe the population of the least liveable cities.
Examining, analysing, interpreting

2. Draw up a table or use a spreadsheet to collect at least five sets of information to compare the top ten and bottom ten in the liveable cities ranking. Use the population data you collected for the previous question as your first set of information. Other possible data sets are number of universities, number of hospitals, population density, any recent violence, traffic issues, the availability of public transport, housing types, presence of slums and water supply and sanitation. Comment on the differences between the most liveable and least liveable cities. Write at least three sentences. **Classifying, organising, constructing**

8.3 EXERCISES

Geographical skills key: GS1 Remembering and understanding **GS2** Describing and explaining **GS3** Comparing and contrasting **GS4** Classifying, organising, constructing **GS5** Examining, analysing, interpreting **GS6** Evaluating, predicting, proposing

8.3 Exercise 1: Check your understanding

1. **GS1** How many cities are ranked in the EIU liveability ranking?
2. **GS1** What is the difference between the score of the top city (Vienna) and the tenth city?
3. **GS1** Name the three lowest ranked cities in the 2018 liveability ranking.
4. **GS1** In which type of climatic region are most of the liveable cities?
5. **GS5** Analyse the information in **FIGURE 1**.
   (a) How many of the top ten most liveable cities are found on each continent?
   (b) How many of the most liveable cities are found in the northern hemisphere?
   (c) Describe the distribution of the least liveable cities in the world.
   (d) How many of the least liveable cities are found on each continent?
   (e) How many of the least liveable cities are found in each hemisphere?

8.3 Exercise 2: Apply your understanding

1. **GS5** London and New York have a similar ranking. Why do you think these well-known cities are ranked so low?
2. **GS5** Why might a city suddenly fall down the liveability rankings?
3. **GS6** What do you think could be done to improve a city's liveability ranking?
4. **GS5** Suggest some reasons why Vienna might have taken the number one ranking in 2018 after sharing it with both Melbourne and Vancouver for several years.

▶

5. **GS2** Why do cities in both Australia and Canada outperform cities in the United States in the liveability rankings?
6. **GS4** Make a list of what you think are the ten most important criteria to judge a liveable city.

Try these questions in learnON for instant, corrective feedback. Go to www.jacplus.com.au.

8.4 SkillBuilder: Drawing a climate graph

What are climate graphs?

Climate graphs, or climographs, are graphs that show climate data for a particular place over a 12-month period. They combine a column graph and a line graph. The line graph shows average monthly temperature, and the column graph shows average monthly precipitation (rainfall).

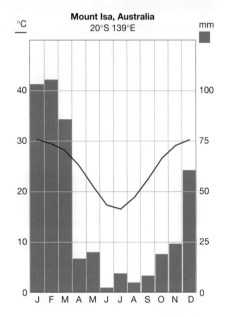

Mount Isa, Australia
20°S 139°E

Select your learnON format to access:

- an overview of the skill and its application in Geography (Tell me)
- a video and a step-by-step process to explain the skill (Show me)
- an activity and interactivity for you to practise the skill (Let me do it)
- questions to consolidate your understanding of the skill.

on Resources

Video eLesson	Drawing a climate graph (eles-1644)	
Interactivity	Drawing a climate graph (int-3140)	
Weblinks	World climate	
	Bureau of Meteorology	

8.5 Melbourne — a liveable city

8.5.1 What is Melbourne like?

What makes Melbourne such a liveable city? Use the **Melbourne view** weblink in the Resources tab to listen to why it was voted one of the world's most liveable cities. Think about whether or not Melbourne is a 'perfect' city. Saying that a city is one of the 'most liveable' gives it a ranking, like those given to cars or restaurants. However, the criteria that are used for the selection process may change. In 2018 Melbourne was ranked as the second most liveable city in the world, behind Vienna. Why?

FIGURE 1 The balance of good features and not-so-good features

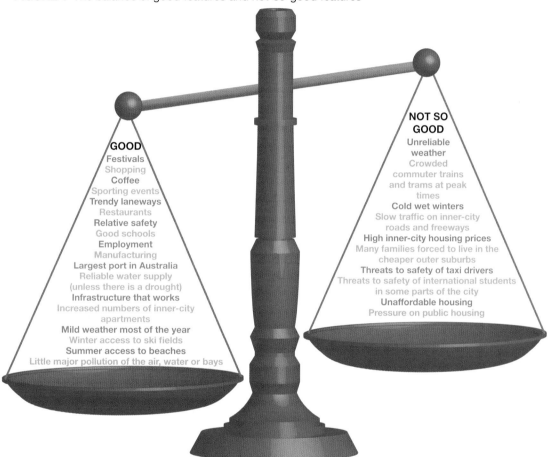

GOOD
Festivals
Shopping
Coffee
Sporting events
Trendy laneways
Restaurants
Relative safety
Good schools
Employment
Manufacturing
Largest port in Australia
Reliable water supply
(unless there is a drought)
Infrastructure that works
Increased numbers of inner-city
apartments
Mild weather most of the year
Winter access to ski fields
Summer access to beaches
Little major pollution of the air, water or bays

NOT SO GOOD
Unreliable weather
Crowded commuter trains
and trams at peak times
Cold wet winters
Slow traffic on inner-city
roads and freeways
High inner-city housing prices
Many families forced to live in the
cheaper outer suburbs
Threats to safety of taxi drivers
Threats to safety of international students
in some parts of the city
Unaffordable housing
Pressure on public housing

8.5.2 Is all of Melbourne the same?

Within the most liveable city, there are some parts that are more liveable than others. People may have higher incomes, larger houses, more and better cars, a view of the sea or the Yarra River, and better shops and entertainment facilities. This does not mean that the highly liveable inner eastern and southern suburbs do not have pockets of poorer housing and homelessness; it is just that they have fewer of them.

Like most large cities, Melbourne is known for its regions. There is a north–south divide, with the Yarra River acting as the boundary, and an east–west divide, with the CBD as its boundary. The west is predominantly flat, dry land, with cheaper housing and land values. The south and east are the leafy suburbs with bayside and hill views, more expensive land and larger houses. The inner north is quite trendy, and more expensive than the west. But the outer suburbs, in all directions, are less well serviced by infrastructure. These areas have huge new suburbs full of large houses on smallish blocks, which are much cheaper than anything close to the centre of the city.

FIGURE 2 Melbourne — a liveable city?

Ⓐ Central Melbourne — the Yarra River, parkland, sporting facilities and central business district

Ⓑ Melbourne tram

Ⓒ Modern office buildings

Ⓓ Graffiti in Hosier Lane

Ⓔ Outer suburban area of South Morang

Ⓕ Studley Park, not far from the city centre

FIGURE 3 Promotion for Melbourne

DISCUSS

Cultural diversity in a place can often be reflected in food, restaurants and festivals.

Use references such as food guides and restaurant phone lists for a major city to identify and record the diversity of restaurants reflecting different cultures. Are there geographical patterns to the locations of restaurants by culture or place? How does such cultural diversity have an impact on the liveability of a place?

[Intercultural Capability]

 Resources

🔗 **Weblink** Melbourne view

8.5 EXERCISES

Geographical skills key: GS1 Remembering and understanding **GS2** Describing and explaining **GS3** Comparing and contrasting **GS4** Classifying, organising, constructing **GS5** Examining, analysing, interpreting **GS6** Evaluating, predicting, proposing

8.5 Exercise 1: Check your understanding

1. **GS2** What are two ways of describing the weather in Melbourne?
2. **GS1** What is the city's traffic like during peak times?
3. **GS1** What is the difference between the landforms in Melbourne's east and west?
4. **GS1** What is the difference in rainfall between Melbourne's east and west?
5. **GS5** Look at the images of Melbourne in **FIGURE 2**. List ten liveability factors that these images illustrate.

8.5 Exercise 2: Apply your understanding

1. **GS6** Which of Melbourne's not-so-good features may have an impact on whether it remains the world's most liveable city?
2. **GS6** Why do you think that four of Australia's major cities are ranked in the world's top ten?
3. **GS3** Explain how Melbourne's weather can be both a positive and a negative in terms of liveability.
4. **GS5** In many parts of Melbourne high-rise apartments rather than single dwellings on a large block are becoming more common. Suggest reasons for this trend.
5. **GS5** Explain why Melbourne has a high dependence on private cars and how this affects liveability.

Try these questions in learnON for instant, corrective feedback. Go to www.jacplus.com.au.

8.6 Liveability and sustainable living

8.6.1 Sustainability

Australia's major cities consistently rate among the most liveable. Liveability, however, is not always the same as sustainability. Sustainability considers how well a community is currently meeting the needs and expectations of its population and how well it will be able to continue providing for its population in the future.

Indicators that a place is sustainable include:
- low working hours to meet basic needs
- easy access to education
- satisfactory and affordable housing
- plenty of recycling and composting
- reliable transport
- low emissions and high air quality
- **biodiversity**
- high renewable energy use and low non-renewable energy use
- good water, forests and marine health
- ability to respond to disasters.

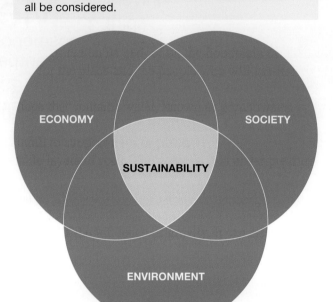

FIGURE 1 To achieve sustainability, a city's environmental, economic and social aspects must all be considered.

Sustainable cities index

This annual index considers 50 leading cities and ranks each against a range of indicators. These are organised under the headings of people (society), planet (environment) and profit (economy).

TABLE 1 Top ten sustainable cities by indicator 2018

Ranking	People	Planet	Profit
1	Edinburgh	Stockholm	Singapore
2	London	Frankfurt	London
3	Paris	Zurich	Hong Kong
4	Taipei	Vienna	New York
5	Stockholm	Copenhagen	Munich
6	Prague	Oslo	Edinburgh
7	Seoul	Hamburg	San Francisco
8	Amsterdam	Berlin	Boston
9	San Francisco	Munich	Zurich
10	Madrid	Montreal	Seoul

Ecological footprint

Everything we do and consume has an impact on the environment. Land is cleared to grow plants and animals; fish are caught in the sea; water is diverted for homes, businesses and farms; and most transport is powered by non-renewable resources. An **ecological footprint** calculates the land area (hectares) that would be needed to sustain an individual (expressed as per capita). It is used to compare the amount of various resources used per capita in countries around the world.

Generally, if you live in a high-income country such as Australia, you are likely to have an ecological footprint that is much larger than a person who lives in a low-income country such as Chad. The average ecological footprint of all people on Earth is 2.84 hectares. The average Australian footprint is about 8.8 hectares. To enjoy a sustainable way of life, the population needs to stay within the Earth's carrying capacity, and the average footprint should not be more than 1.7 hectares. **FIGURE 2** shows that developed countries such as Luxemburg, the United States and Australia far exceed this figure. Australia, for example, is using resources and generating waste more than five times higher than the Earth can regenerate and absorb. As more countries develop industries and improve their standard of living, clever responses will be needed to ensure that everyone can enjoy a high standard of liveability.

FIGURE 2 Top ten countries with the biggest and smallest ecological footprints (hectares per capita) per person, 2013

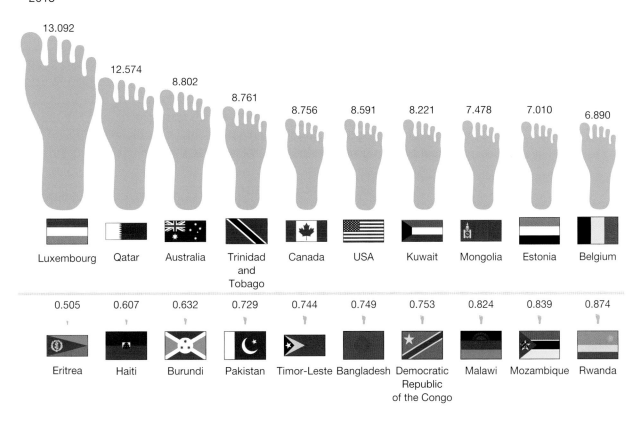

Government policy can influence the ecological footprint through power generation, transport, water, industry support, rubbish collection and building regulations. Individuals can influence the ecological footprint through what they eat and buy, how they use water and power, whether they recycle and compost, and how they build their houses and travel.

DISCUSS

How important and ethical is it to include all aspects of sustainability (environmental, social and economic) when classifying a place and its level of liveability? **[Ethical Capability]**

8.6 INQUIRY ACTIVITY

Find one image that shows living conditions in a country with an ecological footprint of over seven hectares per capita and one image that shows living conditions in a country with an ecological footprint of less than one hectare per capita. Refer to **FIGURE 2** for possible examples.

Annotate your images to explain how the living conditions have an impact on the ecological footprint.

Comparing and contrasting

8.6 EXERCISES

Geographical skills key: GS1 Remembering and understanding **GS2** Describing and explaining **GS3** Comparing and contrasting **GS4** Classifying, organising, constructing **GS5** Examining, analysing, interpreting **GS6** Evaluating, predicting, proposing

8.6 Exercise 1: Check your understanding

1. **GS1** What are the three aspects that are considered in a definition of **sustainability**?
2. **GS1** List three indicators that a place is sustainable.
3. **GS1** Define the term 'biodiversity'.
4. **GS5** Refer to **TABLE 1** and your atlas. Answer the following questions.
   (a) There are 17 cities in the table. How many are located in the continent of Europe?
   (b) Which other continents are represented?
   (c) Which cities are in the top ten for each of the three indicators for a sustainable city?
5. **GS5** Refer to **FIGURE 2** and locate and describe the distribution of countries with an ecological footprint of seven or more hectares per capita. Refer to pattern, directions, continents and latitude.

8.6 Exercise 2: Apply your understanding

1. **GS6** What do you think will happen to the global ecological footprint if liveability improves on every continent?
2. **GS4** Refer to the list of things in section 8.6.1 which indicate that a **place** is **sustainable**. Categorise each indicator as applying to society, economy or **environment**. Suggest one more possible indicator for each category.
3. **GS6** Consider the ways in which resources have been used to improve the liveability in your area. Which aspects would you be prepared to **change** a little so that others might improve the liveability of where they live?
4. **GS2** Explain why high-income countries have a much larger ecological footprint than low-income countries.
5. **GS5** Explain the term 'carrying capacity' and compare the Earth's carrying capacity to its current ecological footprint.

Try these questions in learnON for instant, corrective feedback. Go to www.jacplus.com.au.

8.7 Less liveable cities

8.7.1 Port Moresby

The United Nations measures people's quality of life using the Human Development Index (HDI). In 2000, Papua New Guinea was ranked 133 in the world; in 2018 its ranking had dropped to 153 (out of 189). Its largest city, Port Moresby, faces many challenges to meet the needs of its people and improve the standard of living.

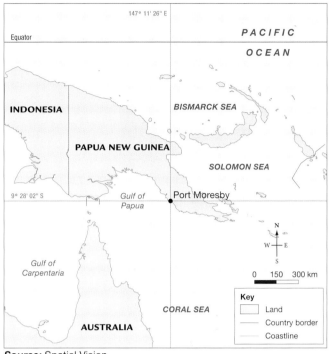

FIGURE 1 Location map of Port Moresby

Source: Spatial Vision

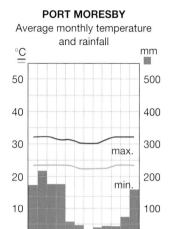

FIGURE 2 Climate graph for Port Moresby

PORT MORESBY
Average monthly temperature and rainfall

Environment

Port Moresby, the capital of Papua New Guinea (PNG), is located on the south-eastern coastline. Its population is approximately 374 500, and it is the largest city in PNG.

Safety

The crime rate in Port Moresby is very high, and the city has a reputation as one of the most dangerous in the world. Crimes are often very violent, and gang-based crime is common. There are not enough police, and many crimes are never solved. Travellers are advised to be very careful, to not wear obviously expensive jewellery and to avoid travelling at night.

Health

The government in PNG spends little on preventative measures such as clean water. It also spends little on healthcare. For instance, not all pregnant women can give birth in a hospital, which leads to many complications in childbirth.

Education

School facilities in PNG are quite poor, and attendance rates are very low, particularly for girls. Poor bus services, lack of interest and inability to pay school fees all influence the attendance rate. Only a small proportion of students complete Year 12. The **literacy rate** of 64.2 per cent is quite low by world standards.

Economy

The government in PNG applies a social security tax to both companies and employees, which is used to fund healthcare and welfare benefits. Unemployment rates are very high and most work is found in the **informal sector**. Many businesses in this sector involve selling food and other goods. About 40 per cent of the population lives on less than $1.25 a day. Fortunately, many families can take advantage of the good growing conditions to produce food to eat and sell.

Life is difficult for girls, and there is much discrimination. Girls do not all get access to school; their literacy rate is lower than that of boys; child-bearing begins at a young age; and the level of violence against women is among the highest in the world.

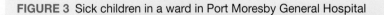

FIGURE 3 Sick children in a ward in Port Moresby General Hospital

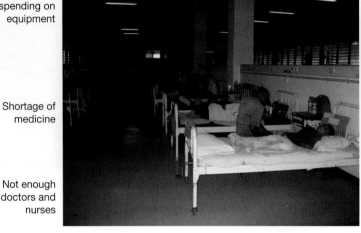

Low spending on equipment

Shortage of medicine

Not enough doctors and nurses

Treatable diseases are common.

Average life expectancy is about 65 years.

Highest HIV/AIDS infection rate in the Pacific region

Infrastructure

FIGURE 4 Port Moresby is a mixture of high-rise urbanised landscapes and village landscapes.

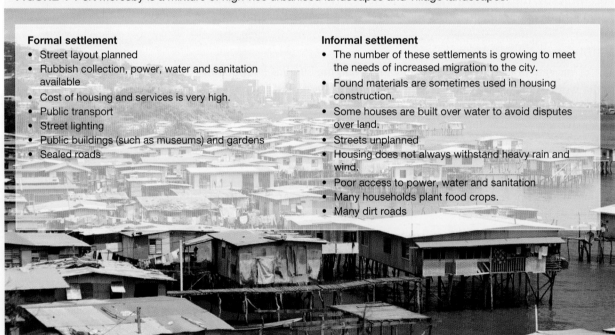

Formal settlement
- Street layout planned
- Rubbish collection, power, water and sanitation available
- Cost of housing and services is very high.
- Public transport
- Street lighting
- Public buildings (such as museums) and gardens
- Sealed roads

Informal settlement
- The number of these settlements is growing to meet the needs of increased migration to the city.
- Found materials are sometimes used in housing construction.
- Some houses are built over water to avoid disputes over land.
- Streets unplanned
- Housing does not always withstand heavy rain and wind.
- Poor access to power, water and sanitation
- Many households plant food crops.
- Many dirt roads

8.7.2 Dhaka — a less liveable city?

Dhaka is the capital city of Bangladesh. Some regions of Dhaka are similar to Australian suburban areas, with solid housing structures, shopping centres, high car ownership, and high expenditure on cars, household possessions, personal services and technology. However, it is the incidence of poverty and unplanned urban growth that leads to the city being ranked as one of the least liveable in the world.

Environment

Dhaka is located in Asia at latitude 23.43°N.

As you can see in **FIGURE 5**:

- there is a distinct dry season
- eighty per cent of rain falls in the wet season (the monsoon)
- it is often hot and humid
- approximately 2000 mm of rain falls per year
- it is warm to hot all year.

Dhaka is only 2–13 metres above sea level. Snow-melt from the Himalayas feeds the rivers. This area is at high risk from climate change because increase in snow- and ice-melt or rainfall will add to river flow. There are many rivers that flow near Dhaka. There is a high risk of flooding. During the monsoon, there are often strong winds, which also cause damage.

FIGURE 5 Climate graph for Dhaka

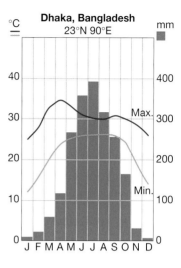

FIGURE 6 Location of Dhaka

Source: Spatial Vision

TABLE 1 Humidity levels in Dhaka

	Jan.	Feb.	Mar.	Apr.	Ma.	Jun.	Jul.	Aug.	Sep.	Oct.	Nov.	Dec.
Humidity (%)	54	50	45	56	72	80	80	79	79	73	67	64

FIGURE 7 Map showing many rivers that flow through and around Dhaka

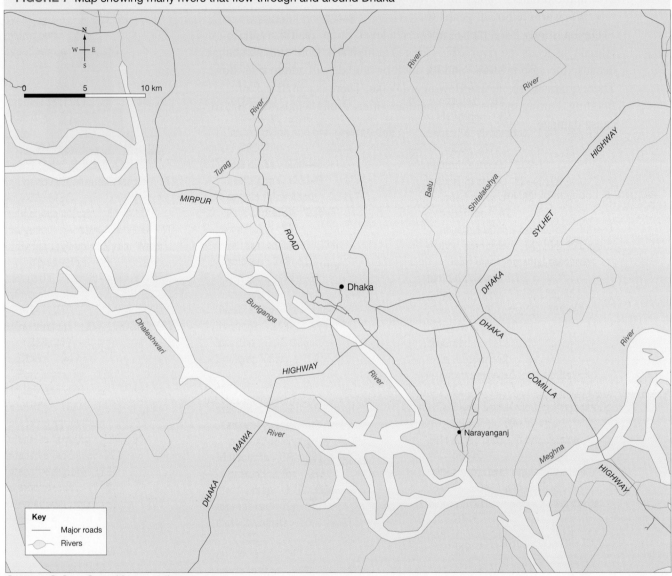

Source: © OpenStreetMap contributors

Infrastructure

The population of Dhaka is more than 18 million, and it is one of the most densely populated cities in the world. Dhaka's population has an annual growth rate of around 4.2 per cent. People migrate to the city in the hope of finding work in the growing industrial sector. The huge influx of people has led to unplanned urban growth on vacant land, and about half the population live in slums. This has created one of the most densely populated cities in the world. Because people can be evicted from the slum areas by landowners, the government does not provide infrastructure.

FIGURE 8 A communal water pump in a slum region

Water is not pumped to houses.

Houses are built from found materials.

Some people have to walk 900 metres to a toilet.

Gangsters control the cost and supply of electricity.

Roads are poor and congested.

There are no gutters or sewerage systems.

Safety

Crime rates are high in the poor areas of Dhaka. There is gangster violence; land grabbing; violence against women and children; arson; and crimes related to gambling, drugs, alcohol and illegal weapons. There are not enough police officers, and they cannot be relied on to protect citizens.

Education

Primary education is compulsory, but the government is unable to provide enough schools and resources for the increasing population. Many students do not attend school all the time because their families need them to earn money. In spite of the tough conditions, the education rate in the city is slightly higher than in rural areas, and the national literacy rate is about 73 per cent.

Economy

Most jobs are found in the informal sector; examples include rickshaw driver, street vendor and garment worker. Women are excluded from trades and transport, and most find work as servants or in agriculture. (Food is grown on vacant land within and around the city.) The pay in these types of jobs is low, and most or all household members need to work.

FIGURE 9 This woman has to walk through floodwaters to collect drinking water. Poor areas have no drainage, and floodwater quickly spreads into houses and over paths.

FIGURE 10 Employment by sector in Bangladesh

Services 39.84%

Industry 21.09%

Agriculture 39.07

FIGURE 11 Children as young as seven undertake exhausting work. This child is earning $1 a day.

Child labour is common; It is estimated that 8 per cent of children under the age of ten are working, predominantly in the manufacturing sector. Between the ages of 8 and 14 it is estimated that almost 50 per cent of children had full time jobs. Children as young as six have been found working instead of attending school. In 2016 about 24.3 per cent of the population lived below the poverty line, surviving on less than $1.90 per day. This is an improvement on the 2010 rate, when 31.5 per cent of the population was classed as living below the poverty line. Even with these disadvantages, many people think the city offers a better quality of life than the rural areas do.

Healthcare

Healthcare is mainly provided through hospitals, which are located in the **formal** part of the city. There is a shortage of hospital beds, equipment, doctors and nurses. There are no medical facilities in the slums, and often families cannot afford to pay for treatment. Private charity groups do offer some programs, particularly for maternal health.

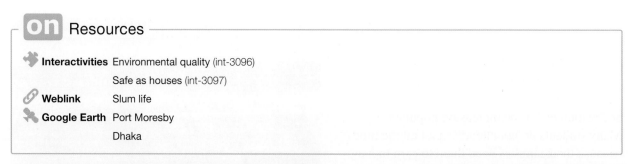

on Resources

Interactivities Environmental quality (int-3096)

Safe as houses (int-3097)

Weblink Slum life

Google Earth Port Moresby

Dhaka

8.7 INQUIRY ACTIVITIES

1. Find an image of an informal settlement in a country other than PNG, such as the favelas in Rio, Brazil. What are the advantages and disadvantages of informal settlements? **Examining, analysing, interpreting**

2. The population density of Dhaka is approximately 40 000 people per square kilometre. Select one square kilometre of a residential region near your school. Estimate the number of people who live in that **space**. **Evaluating, predicting, proposing**

3. Choose one of the other least liveable cities (see subtopic 8.3) and find out how the natural environment creates challenges and provides benefits there. Consider the city's location, climate and landscape. **Examining, analysing, interpreting**

4. Use the **Slum life** weblink in the Resources tab to watch a video showing life in slums. Explain how slums are part of the process of city growth. **Describing and explaining**

8.7 EXERCISES

Geographical skills key: GS1 Remembering and understanding **GS2** Describing and explaining **GS3** Comparing and contrasting **GS4** Classifying, organising, constructing **GS5** Examining, analysing, interpreting **GS6** Evaluating, predicting, proposing

8.7 Exercise 1: Check your understanding

1. **GS1** Refer to **FIGURE 1**. At what latitude is Port Moresby?
2. **GS1** Why don't all children attend school in Port Moresby?
3. **GS1** In which sector of the economy do most people find work?
4. **GS2** How does environmental quality (such as climate) influence living conditions in Port Moresby?
5. **GS2** Why are travellers advised to be careful in Port Moresby?
6. **GS5** Refer to **FIGURE 3**. Which is the biggest health issue facing Port Moresby? Why?
7. **GS1** In which continent is Dhaka?
8. **GS1** When is the monsoon season in Dhaka?
9. **GS2** How does the natural *environment* influence life in Dhaka?

8.7 Exercise 2: Apply your understanding

1. **GS3** What is the difference between the population of Port Moresby and the biggest city in your state or territory?
2. **GS3**
   (a) Compare the literacy rate in PNG and Australia.
   (b) Compare the life expectancy in PNG and Australia.
   (c) Compare the HDI ranking of PNG and Australia.
3. **GS3** Compare the population of Dhaka with the population of Australia.
4. **GS5** How is life in the Dhaka slums affected by the lack of resources that are normally provided by government, such as water, healthcare, education and safety?
5. **GS2** Why do very young children enter the workforce?
6. **GS2** Why does Dhaka continue to grow even though it rates poorly in terms of liveability?

Try these questions in learnON for instant, corrective feedback. Go to www.jacplus.com.au.

8.8 Improving liveability

8.8.1 Distribution of hunger

A basic human requirement is food, and access to enough food is a strong measure of liveability. Even in a world where there is plenty of food and millions of people are overweight, about one person in eight does not have enough to eat.

There are approximately 815 million **undernourished** people, or one in ten in the world today. Many children in poorer countries are underweight and do not get enough food to be healthy and active.

Three-quarters of all hungry people live in rural areas, mainly in the villages of Asia and Africa (see **FIGURE 1**). Most of these people depend on **agriculture** for their food. They rarely have other sources of income or employment. As a result, they may be forced to live on one-quarter of the recommended calorie intake and a small amount of water each day.

If enough rain does not fall at the right time of year, crops will not grow well and there will be little grass for **livestock**. However, rainfall is not the only factor contributing to hunger. **FIGURE 2** summarises causes of hunger.

FIGURE 1 Distribution of hunger, 2018

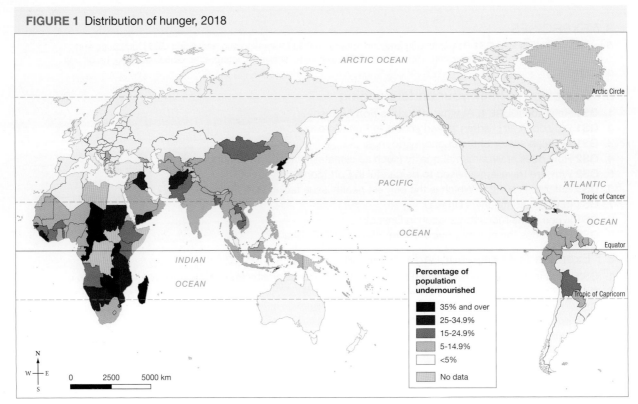

Source: Food and Agriculture Organisation

FIGURE 2 Causes of hunger

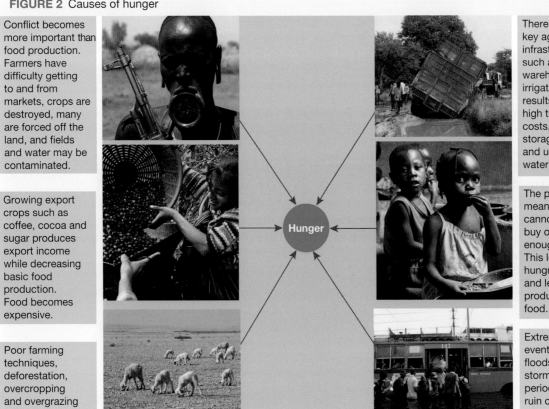

Conflict becomes more important than food production. Farmers have difficulty getting to and from markets, crops are destroyed, many are forced off the land, and fields and water may be contaminated.

Growing export crops such as coffee, cocoa and sugar produces export income while decreasing basic food production. Food becomes expensive.

Poor farming techniques, deforestation, overcropping and overgrazing reduce soil fertility.

There is a lack of key agricultural infrastructure, such as roads, warehouses and irrigation. The results include high transport costs, lack of storage facilities and unreliable water supply.

The poverty cycle means the poor cannot afford to buy or produce enough food. This leaves them hungry and weak and less able to produce more food.

Extreme weather events such as floods, tropical storms and long periods of drought ruin crops and infrastructure.

Hunger

8.8.2 Impact of hunger

A lack of energy and poor health caused by a lack of food are made even worse by poor nutrition.

TABLE 1 The impact of hunger is felt by individuals, families, communities, regions and whole countries.

Social impacts	Economic impacts	Environmental impacts
People become unwell.	Food production declines.	Soil is overused.
Many people (particularly children) die.	The population of cities grows.	Too much land is cleared.
Fathers leave in search of work.	Poverty increases.	Soil fertility and local biodiversity decline.
There is political unrest.	The government cannot afford new infrastructure.	

8.8.3 Ending hunger

There is a range of organisations that focus on reducing hunger. Sometimes food is provided for immediate consumption and sometimes projects are undertaken to increase food production in the future. Actions can happen on a range of scales:

- Individuals in any country can join groups or donate to organisations that work to reduce hunger.
- The government of the affected country can provide assistance to the poor or improve infrastructure.
- Other countries can provide financial and food aid or consider the impact of their own policies.

8.8.4 Sustainable Development Goals

Many countries cannot afford to provide infrastructure for their growing population. The underlying cause of very low liveability is poverty. Reducing poverty is fundamental to improving living conditions in many parts of the world.

United Nations Development Goals

The United Nations (UN) is an organisation with members from 193 countries. In 2000, 189 countries signed a pledge to free people from extreme poverty by 2015 (Millennium Development Goals 2000–2015). In 2015, a new pledge was signed with 17 goals, each with specific targets to be reached over 15 years (Sustainable Development Goals 2015–2030).

TABLE 2 UN Development Goals

Millennium Development Goals 2000–2015	Examples of achievements of MDGs	Sustainable Development Goals 2015–2030	
Eradicate extreme poverty and hunger	Fewer people live in extreme poverty.	No poverty	Industry, innovation and infrastructure
Achieve universal primary education	Primary school enrolments have increased.	Zero hunger	Reduce inequality
Promote gender equality and empower women	Many more girls are attending school.	Good health and wellbeing	Sustainable cities and communities
Reduce child mortality	More babies are surviving.	Quality education	Responsible consumption and production
Improve maternal health	More mothers have access to healthcare when giving birth.	Gender equality	Combat climate change
Combat HIV/AIDS, malaria and other diseases	Vaccination has reduced incidence of measles.	Clean water and sanitation	Conserve and use ocean resources sustainably

(continued)

TABLE 2 UN Development Goals *(continued)*

Millennium Development Goals 2000–2015	Examples of achievements of MDGs	Sustainable Development Goals 2015–2030	
Ensure environmental sustainability	Safe water is available to more people.	Affordable and clean energy	Protect and use Earth's resources sustainably
Develop a global partnership for development	Huge increase in number of people with phone and internet	Decent work and economic growth	Provide access to justice and promote peaceful societies

Australian government and NGOs

The Australian government recognises that we are **global citizens**, and it supports an overseas aid program through its Department of Foreign Affairs and Trade. Overseas aid helps improve outcomes in health, education, economic growth and disaster response in many locations.

The Australian government runs projects to improve living conditions, often working with other countries or with **non-government organisations** (NGOs). NGOs also run programs on their own. Well-known NGOs include World Vision, CARE Australia and Australian Red Cross.

Small changes, big results

Simple and **appropriate technology** can make an enormous difference to people's lives in developing countries (see **FIGURE 5**).

FIGURE 3 Countries receiving assistance from Australia

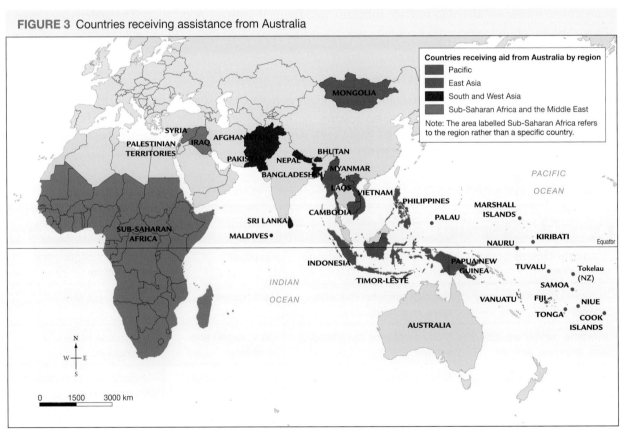

Source: Department of Foreign Affairs and Trade

In addition, a small amount of money can sometimes make a big difference to an individual or community group. Microfinance, or microcredit, is a system of lending small amounts of money, perhaps $150. The money is used to invest in something that can generate income. A person might buy an animal for milking and breeding, equipment for basket-making, stock for a store, or materials for jewellery-making. The loan must be repaid, but at a low interest rate, and further loans can be taken out.

FIGURE 4 Examples of projects to improve liveability (a) A child immunisation clinic on the Kokoda Track (b) Building schools and improving education in Indonesia (c) Planting grasses in Fiji to stabilise sea banks.

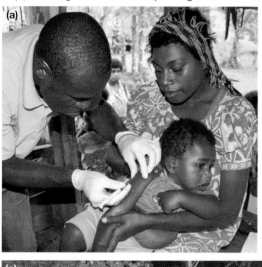

FIGURE 5 Appropriate technology (a) Electricity in Nepal is not available to all houses, so a solar lamp increases the opportunities to read. (b) In South Africa, people push hippo rollers, which make it easier to collect water from distant wells and bring it home.

8.8 INQUIRY ACTIVITIES

1. Work with a partner to find an example of a project that is trying to solve the immediate issue of hunger and an example of a project that is trying to make food production *sustainable*. Describe where the project is taking place and which organisation manages the project. Create an outline of the project. Refer to **FIGURE 2** and explain which of the causes of hunger will be reduced by the project.
Evaluating, predicting, proposing

2. Prepare a report about the work of one NGO involved in programs that aim to improve liveability in an overseas country. Include background information about the location of the project (country and locality); statistical data about living conditions (such as life expectancy, access to safe water, doctors per 100 000 people); and *environmental* conditions. Describe the NGO and its project and how it is aiming to improve liveability. **Examining, analysing, interpreting**

3. Choose one of the Sustainable Development Goals. Use a visual organiser to explain how achieving this goal will improve liveability. Consider the flow-on effects and the impact on people, the economy and the *environment*. **Classifying, organising, constructing**

8.8 EXERCISES

Geographical skills key: GS1 Remembering and understanding **GS2** Describing and explaining **GS3** Comparing and contrasting **GS4** Classifying, organising, constructing **GS5** Examining, analysing, interpreting **GS6** Evaluating, predicting, proposing

8.8 Exercise 1: Check your understanding

1. **GS5** Refer to **FIGURE 1**.
   (a) Which region has the largest number of hungry people? Name three countries in this region.
   (b) Describe the *change* in the distribution of hunger. Use the following questions to help.
   - In which regions has the number of hungry people increased?
   - By what percentage?
   - In which regions has the number of hungry people decreased?

2. **GS1** Copy and complete the following sentence to make it accurate. 'Most of the world's hungry people live in _____ villages in _____ and _____.'

3. **GS3** Refer to **FIGURE 1**. In 1990–1992 there were about one billion people who did not get enough food. How many people suffered from hunger in 2018? Is this an increase or decrease? By how many million has it changed?

4. **GS2** How can poor roads contribute to hunger?

5. **GS1** Which organisation developed the Millennium Development Goals and the Sustainable Development Goals?

6. **GS1** How many countries are in there in the world? What percentage of countries supported these sets of goals?

8.9 Liveable communities and me

8.9.1 Liveability studies

A study of a region's liveability will reflect its natural characteristics and human characteristics. All communities would like a safe, healthy and pleasant place to live, a sustainable environment, the chance to earn a liveable wage, reliable infrastructure and opportunities for social interaction.

The findings of a liveability survey will be influenced by a range of factors.

- Where a person lives influences their access to services, employment and environmental features, and their address may influence their perception of the quality of the region.
- Different age groups have different views and needs.
- Current economic conditions influence a person; for example, a major employer may have closed or opened.
- Environmental conditions affect a person; for example, a region may be experiencing drought.
- Government policies influence infrastructure, housing assistance and grants to local sports clubs.

To find out about the liveability of an area, a number of themes need to be investigated. Some of these can be gained from **census** statistics, while others can be gained only through surveys and fieldwork.

TABLE 1 Matching liveability indicators to key themes

Measure	Examples of indicators	
Social	• Population characteristics (gender, age) • Education (primary, secondary, tertiary) • Health (life expectancy, health-centre attendance, length of walking tracks, smoking rates, weight, chronic diseases) • Safety (perception, crime rates, road deaths and injuries, work safety)	• Volunteering • Voting • Aged care accommodation • Access to public transport • Membership of clubs and organisations • Diversity (ethnicity)
Environmental	• Biodiversity • Planning for the future • Water access • Waste management • Ecological footprint	• Public spaces • Household recycling • Weather • Land clearing
Economic	• Employment • Variety of businesses • Income • Financial stress • Housing types	• House ownership • Infrastructure • Internet access • Power • Car ownership

In any community there will usually be agreement about some things that improve liveability. All groups accept that safe water, sealed roads and a reliable power supply are important.

If a community wants to obtain certain kinds of items on its liveability 'wish list', it sometimes needs help from national, state or local government. Examples of such items include major roads, railways and desalination plants. Sometimes, though, a wish-list item is best obtained by an individual or community. This is the case when setting up sporting clubs, youth groups and local music events.

Community wish list

- Playgrounds
- Paths for prams
- Primary schools
- Single-person housing
- Family housing
- Friendly community
- Shopping nearby
- Paths for scooters
- Health services
- Public transport
- Neighbourhood house
- Parks and gardens
- Public seating
- Recognition of those from non-English-speaking backgrounds
- Financial security
- University of the Third Age

FIGURE 1 Community wish list: some aspects of liveability are common to all groups and some are desired by particular groups.

FIELDWORK TASK

Looking at your school environment

Geographers are particularly interested in:
- the location of things
- the distance between things
- the distribution patterns we can see when we produce a map
- the movement between places
- the connection between places
- the changes that happen over time.

How can you apply these concepts when finding ways of improving your school environment? Work in pairs to gather the data needed for the following fieldwork task. Each student will complete a report of findings.

FIGURE 2 A modern school environment

Step 1: Study the distribution of resources and landscapes over space in the schoolyard.
- Obtain an outline map of the schoolyard.
- Walk around the schoolyard and identify different categories of land use. Design a key for your map and mark in the land uses on your map.
- As you walk around the schoolyard, also note the landscape (slopes, swamp, bare ground, concrete and so on). Record the information about the landscape on your map. You may add to your key or annotate the map. You may also wish to use an overlay. (See the SkillBuilder in subtopic 8.10.) You could also add images.
- Describe the distribution of land uses in the schoolyard. Can you identify regions? Mark these on your map. Consider using an overlay. Is there any interconnection between the landscape and land use?

Step 2: Study the patterns of movement in the schoolyard.
- Choose two places that students often walk past.
- Record the number of students who pass and the direction of travel in two 5-minute sessions.
- Add this information to your map. Which are the busiest walkways in the schoolyard?

Step 3: Make recommendations for improvements to the schoolyard.
Based on the information you have gathered, describe:
- pleasing aspects of the schoolyard
- disappointing aspects of the schoolyard
- your three most effective suggestions for improving the schoolyard.

Mark the location of the proposed changes on your map. You may include images of proposed changes.

Conduct a short survey of ten other students to find out what they think about your three suggestions for improving the schoolyard.

When conducting your survey, there are three types of questions you need to ask: those that seek an opinion, those that seek a fact and those that seek a suggestion.
- To seek opinions, try to ask for a rating. For example, There should be more seats in the schoolyard. Do you (a) strongly disagree? (b) disagree? (c) neither agree nor disagree? (d) agree? (e) strongly agree?
- To seek facts, ask for a structured response. For example, How often do you use the seating by the oval? (a) Every day (b) 1–3 times a week (c) A few times a month (d) A few times a year (e) Almost never.
- To seek suggestions, use open-ended questions. For example, What sorts of plants do you think should be planted along the front fence?

Create one question for each of your suggestions. Cover each type of question. Ensure you have your map with you when you ask the questions.

Collate, or gather, the survey data for each question. For each question, count the number of responses that are the same, and present the result in words, in a table or in a graph. (You might choose a bar graph, for example). There might be a clear trend in the responses or there might be variety in the responses.

To what extent do your survey findings support your suggestions? Describe the result.

8.9.2 Transport strategies

People in towns and cities are always looking for strategies to improve their living conditions. A community is made up of people from a range of age groups, a number of different land uses, a range of needs and a variety of interests. Ideas and plans for improvement may be overarching or targeted.

The movement of people within and between neighbourhoods is an important issue in towns and cities. The humble bicycle is now seen as a way of increasing mobility, reducing traffic congestion, reducing air pollution and boosting health. Bicycle tracks encourage recreational riding for all ages (see **FIGURE 3**) and dedicated bicycle paths along main routes (see **FIGURE 4**) encourage people to commute by bicycle, rather than car, to work and school.

In 1965, a group in Amsterdam, the Netherlands, introduced the idea of bike sharing — public bicycles that are hired, usually for short trips. This first attempt was not a success, but the idea persisted. Modern bike-sharing systems have overcome problems of theft and vandalism by using easily identifiable specialty bicycles, monitoring the bicycles' locations with radio frequency or GPS, and requiring credit-card payment or smart-card-based membership to check-out bicycles. In some places, bicycles can be located on your mobile phone, and there are more links between bicycles and existing public transport. Between 2014 and

2018 bike-sharing programs doubled in size. More than 1600 programs are in operation, providing almost 18.2 million bikes to 20 million registered users.

FIGURE 3 Recreational riding along a trail mainly designed for bicycles

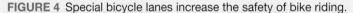

FIGURE 4 Special bicycle lanes increase the safety of bike riding.

Copenhagen was rated as the world's most bike-friendly city in 2014 and has retained this position every year since. Beijing is the world leader in bike-share programs, with 2.4 million share bikes and 11 million registered users. Bike-sharing programs are an example of a popular strategy that is aimed at improving liveability for a range of ages and locations within a community.

An example of a successful bike-sharing scheme is in Paris. The Vélib was introduced in 2007 and quickly doubled in size. By 2012, bicycle trips in the city had grown by 41 per cent. The program continues to grow; an additional 20 000 bikes will be added and the number of share stations expanded to 1400. One-third of the new bikes will have an electric motor with a range of around 50 kilometres when fully charged. These new bikes will also feature a basket with a carrying capacity of 50 kilograms.

It is anticipated that these new bikes will overcome problems associated with maintaining a share-bike program in hilly or uneven terrain, where commuters will ride a bike downhill in the morning, but then elect to return home using alternative transport; leading to a surplus of bikes in one area and a lack of them in others. Bike sharing is part of a plan to reduce car traffic and pollution in Paris, which includes closing streets to cars on weekends, reducing speed limits, encouraging bus travel and extending bicycle lanes.

FIGURE 5 Bike sharing is on the rise globally.

Bike sharing clicks into a higher gear
Estimated number of bike-sharing programs in operation worldwide

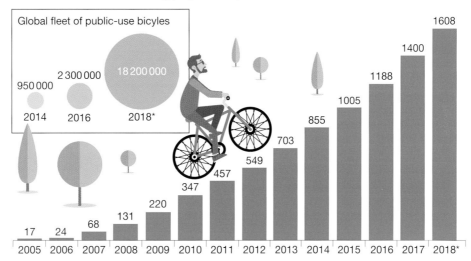

The global rise of bike sharing

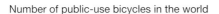

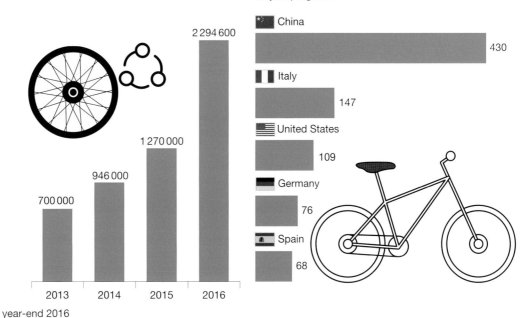

Number of public-use bicycles in the world

Top five countries by number of public-use bicycle programs*

* year-end 2016

@StatistaCharts

Source: Russel Meddin, Bike-sharing Blog

RESEARCH TASK

Teenage spaces in your community

A community is made up of a number of groups that interconnect. Teenagers are an important part of any community.

1. Produce a pie graph to show the population of your community. Refer to the latest census data provided by the Australian Bureau of Statistics. Show the percentage of each of the following categories: less than 13 years old, teenagers, adults and elderly.
   Participation by teenagers in the community will be influenced by values, abilities and interests.

2. (a) Find and read a news article in the local media that is about teenagers. Note the source and date and summarise it in three dot points. For example, is it about an issue that relates to only teenagers, a space used by teenagers, an achievement by teenagers, a complaint about teenagers or a positive story about teenagers?
   (b) As a class or in a small group, brainstorm a list of the ways in which teenagers participate in the community. Divide the agreed list into the following categories: informal, formal, social, cultural and physical.

3. (a) Find a map of your local area and use dots to show the spaces that are most attractive to teenagers. Ensure that your map satisfies all mapping conventions (BOLTSS).
   (b) Describe the distribution pattern of attractive places. Is the pattern linear (in a line or lines), clustered (in small groups) or scattered?
   (c) Think about the pattern you have mapped and your knowledge of the region. Are most of your favourite spaces indoors or outdoors? To what extent is there a connection between the pattern on your map and other features in your neighbourhood? Are your favourite spaces in places that are strongly influenced by the natural environment or the built environment?
   (d) Add an overlay to your **base map** and use dots to show the least favourite spaces for teenagers. Describe the pattern shown on your overlay map. To what extent is there a connection between the pattern on your overlay map and other features in your neighbourhood?

4. Provide up to three examples of ways you participate in communities bigger than the local neighbourhood. For each example:
   (a) Describe the scale of that community. Does it cross local council borders, state borders or national borders?
   (b) Refer to a relevant map (for example, in your atlas or a street directory) to find out the direction and distance from that place to where you live. Add an arrow to your map pointing in the correct direction. Add a label to the arrow to describe the activity and the distance.

Improving your community

5. (a) Identify a space in your neighbourhood that you think could be improved for teenagers. It may be one that is currently attractive, or it may be a least favourite space.
   (b) Provide an image (photograph, diagram or map) of this space. Annotate the image to describe its current characteristics.
   (c) Identify the key concerns about this space. You might think about safety, tolerance, sustainability, access, inclusiveness, services, environmental quality, health and respect.
   (d) How would you improve this space?
      • To help you think of suggestions, use your research skills to find out about ways in which liveability has been improved for teenagers in other parts of the world. Consider European countries in particular.
      • Discuss the ways in which the European ideas are relevant, or not relevant, to your community.
      • Provide a planning suggestion for each of the concerns you raised in question 5(c).
   (e) Provide a new image to show the impact of your proposals. This could be a diagram, sketch, annotated photograph, model or whatever helps communicate what the impacts might be.
   (f) Which are your two most important suggestions? What criteria did you use to choose these suggestions? Which suggestion is most likely to be implemented? Why?
   (g) Compare your suggestions to the ideas of others in your class. What are the common elements? What would you put in a master plan for teenage spaces in your community?

 Resources

8.9 INQUIRY ACTIVITIES

1. Find a local news story about a ***change*** to liveability in your area. Is the ***change*** economic, social or environmental? Is the ***change*** predicted to be positive or negative? Will the ***change*** be permanent?

         Examining, analysing, interpreting

2. Find out about a bike-sharing scheme in Australia or overseas. Describe its location and the region it covers. Provide three other key facts about the scheme.   **Describing and explaining**

3. Some cities provide schemes to encourage people to ride bikes. Find out about the success of bike incentive schemes in European cities. Include the name of the city, the date of the scheme, summary of the scheme and evidence of success or failure.   **Examining, analysing, interpreting**

8.9 EXERCISES

Geographical skills key: GS1 Remembering and understanding **GS2** Describing and explaining **GS3** Comparing and contrasting **GS4** Classifying, organising, constructing **GS5** Examining, analysing, interpreting **GS6** Evaluating, predicting, proposing

8.9 Exercise 1: Check your understanding

1. **GS2** What are the three themes used when investigating liveability? Why do you think these are chosen?
2. **GS4** Refer to **TABLE 1** and identify two aspects that could be placed in a different theme. Justify your suggested ***change***. Suggest one more indicator that should be included. In which theme would it belong?
3. **GS1** What is the name of the official survey used to gather information about people in Australia, and how often is it conducted?
4. **GS1** What are three advantages of increasing bicycle riding?
5. **GS1** What problems were faced by the first bike-sharing scheme?

8.9 Exercise 2: Apply your understanding

1. **GS3** Refer to **FIGURE 1** and use an organiser like a Venn diagram to compare and contrast the liveability wish lists for young families and older people.
2. **GS6** How could the improvement in liveability for one age group actually help the liveability of another age group? Provide an example.
3. **GS6** Refer to **FIGURE 5**.
   (a) Suggest a reason for the rapid increase in the number of share bikes in cities around the world.
   (b) Why do you think China is the fastest growing market for share bikes?
   (c) Copenhagen has been rated the most bike-friendly city in the world but does not appear in the top five countries for bike-share programs. Research the city of Copenhagen and suggest a reason for this.
   (d) How might the introduction of electric bikes encourage more people to use share bikes?
   (e) What problems might the use of electric bikes cause?
4. **GS2** Refer to **FIGURE 1**. Which three items from the community wish list do you think are most needed in the area you live in? Give reasons for your answer.
5. **GS5** Would a bike-sharing scheme be a viable option in your area? Give reasons for your answer.

Try these questions in learnON for instant, corrective feedback. Go to www.jacplus.com.au.

8.10 SkillBuilder: Creating and analysing overlay maps

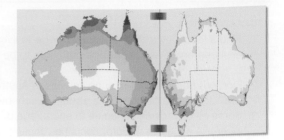

What are overlay maps?

An overlay map usually consists of two or more maps of the same area. A base map is overlaid with a transparent overlay, showing different information. Overlay maps allow users to see the relationships between the information on two or more maps.

Select your learnON format to access:

- an overview of the skill and its application in Geography (Tell me)
- a video and a step-by-step process to explain the skill (Show me)
- an activity and interactivity for you to practise the skill (Let me do it)
- questions to consolidate your understanding of the skill.

on Resources

Video eLesson Creating and analysing overlay maps (eles-1645)
Interactivity Creating and analysing overlay maps (int-3141)

8.11 Thinking Big research project: Liveable cities investigation and oral presentation

SCENARIO

Every year the Economic Intelligence Unit ranks 140 major cities based on five key indicators. These are stability, healthcare, culture and environment, education, and infrastructure. The city that receives the number 1 ranking is the most liveable city in the world; while the city ranked number 140 is the least liveable city.

As part of your investigation you will need to investigate a top ten most liveable city and a top ten least liveable city. You will need to compare them against the key indicators used to rank cities.

Select your learnON format to access:

- the full project scenario
- details of the project task
- resources to guide your project work
- an assessment rubric.

on Resources

 **projectsPLUS** Thinking Big research project: Liveable Cities Investigation and oral Presentation (pro-0239)

8.12 Review

8.12.1 Key knowledge summary

Use this dot point summary to review the content covered in this topic.

8.12.2 Reflection

Reflect on your learning using the activities and resources provided.

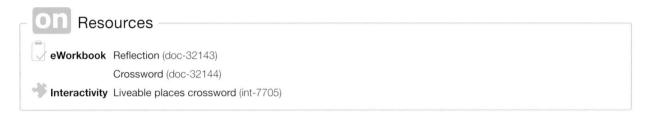

on Resources

eWorkbook Reflection (doc-32143)

Crossword (doc-32144)

Interactivity Liveable places crossword (int-7705)

KEY TERMS

agriculture the cultivation of land, growing of crops or raising of animals

appropriate technology technology designed specifically for the place and the people who will use it. It is affordable and can be repaired locally.

base map the map underneath an overlay

biodiversity the variety of life in the world or in a particular habitat or ecosystem

census a regular survey used to determine the number of people living in Australia. It also has a variety of other statistical purposes.

distribution the way things are spread across an area

ecological footprint the total area of land that is used to produce the goods and services consumed by an individual or country

formal describes an event or venue that is organised or structured

global citizens people who are aware of the wider world, try to understand the values of others, and try to make the world a better place

informal sector jobs that are not officially recognised by the government as official occupations and that are not counted in government statistics

infrastructure the basic physical and organisational structures and facilities that help a community run, including roads, schools, sewage and phone lines

literacy rate the proportion of the population aged over 15 who can read and write

livestock animals raised for food or other products

natural environment elements — such as wind, soil, flowing water, plants and animals — that influence the characteristics of an area

non-government organisation non-profit group run by people (often volunteers) who have a common interest and perform a variety of humanitarian tasks at a local, national or international level

population density the number of people living in a square kilometre

temperate climate climate with generally warm summers and cool winters, without extremes

undernourished not getting enough food for good health and growth

GLOSSARY

adobe bricks made from sand, clay, water and straw and dried by the sun

aerial photograph a photograph taken of the ground from an aeroplane or satellite

agriculture the cultivation of land, growing of crops or raising of animals

alluvial soil a fine-grained fertile soil brought down by a river and deposited on its bed, or on the floodplain or delta

alluvium the loose material brought down by a river and deposited on its bed, or on the floodplain or delta

appropriate technology technology designed specifically for the place and the people who will use it. It is affordable and can be repaired locally.

aquifer a body of permeable rock below the Earth's surface that contains water, known as groundwater. Water can move along an aquifer.

arid lacking moisture; especially having insufficient rainfall to support trees or plants

artesian aquifer an aquifer confined between impermeable layers of rock. The water in it is under pressure and will flow upward through a well or bore.

aspect the direction in which something is facing

atmosphere the layer of gases surrounding the Earth

avalanche rapid movement of snow down a slope, usually under the influence of gravity. It can also be triggered by animals, skiers or explosions.

barometer an instrument used to measure air pressure

base map the map underneath an overlay

biodiversity the variety of life in the world or in a particular habitat or ecosystem

biomass organic (once living) matter used as fuel

blue water the water in freshwater lakes, rivers, wetlands and aquifers

built environment a place that has been constructed or created by people

catchment area the area of land that contributes water to a river and its tributaries

census a regular survey used to determine the number of people living in Australia. It also has a variety of other statistical purposes.

climate change a change in the world's climate. This can be very long term or short term, and can be caused by human activity.

community a group of people who live and work together, and generally share similar values; a group of people living in a particular region

country the place where an Indigenous Australian comes from and where their ancestors lived; it includes the living environment and the landscape.

crevasse a deep crack in ice

cumulonimbus clouds huge, thick clouds that produce electrical storms, heavy rain, strong winds and sometimes tornadoes. They often appear to have an anvil-shaped flat top and can stretch from near the ground to 16 kilometres above the ground.

cyclones intense low pressure systems producing sustained wind speeds in excess of 65 km/h. They develop over tropical waters where surface water temperature is at least 26 °C.

demographic describes statistical characteristics of a population

desalination a process that removes salt from sea water

discharge the volume of water that flows through a river in a given time

distribution the way things are spread across an area

drainage basin the entire area of land that contributes water to a river and its tributaries

drought a long period of time when rainfall received is below average

ecological footprint the total area of land that is used to produce the goods and services consumed by an individual or country

economic relating to wealth or the production of resources

El Niño the reversal (every few years) of the more usual direction of winds and surface currents across the Pacific Ocean. This change causes drought in Australia and heavy rain in South America

elevation height of a place above sea level

evaporate to change liquid, such as water, into a vapour (gas) through heat

evaporation the process by which water is converted from a liquid to a gas and thereby moves from land and surface water into the atmosphere

extensive land use land use in which farms are huge, with few workers and not many cows or sheep per hectare

flood inundation by water, usually when a river overflows its banks and covers surrounding land

fly in, fly out (FIFO) describes workers who fly to work in remote places, work 4-, 8- or 12-day shifts and then fly home

formal describes an event or venue that is organised or structured

fossil fuels fuels that come from the breakdown of living materials, and which are formed in the ground over millions of years. Examples include coal, oil and natural gas.

frostbite damage caused to the skin when it freezes, brought about by exposure to extreme cold. Extremities such as fingers and toes are most at risk, along with exposed parts of the face.

gale force wind wind with speeds of over 62 kilometres per hour

global citizens people who are aware of the wider world, try to understand the values of others, and try to make the world a better place

green water water that is stored in the soil or that stays on top of the soil or in vegetation

groundwater a process in which water moves down from the Earth's surface into the groundwater

hailstone an irregularly shaped ball of frozen precipitation

hailstorm any thunderstorm that produces hailstones large enough to reach the ground

horticulture the growing of garden crops such as fruit, vegetables, herbs and nuts

hydrologic cycle another term for the water cycle

hypothermia a condition in which a person's core body temperature falls below 35 °C and the body is unable to maintain key systems. There is a risk of death without treatment.

improved drinking water drinking water that is safe for human consumption

incentive something that encourages a person to do something

informal sector jobs that are not officially recognised by the government as official occupations and that are not counted in government statistics

infrastructure the basic physical and organisational structures and facilities that help a community run, including roads, schools, sewage and phone lines

intensify to become stronger

intensive agriculture any method of farming that requires concentrated inputs of money and labour on relatively small areas of land; for example, battery hens and rice cultivation

intensive farming farming that uses a lot of resources per hectare and changes the look of the region

intensive land use land use in which farms are smaller but have more workers and machinery to produce high yields per hectare; examples are dairy and poultry farms, orchards, vegetables and feedlots.

inundate to cover with water, especially floodwater

irrigation water provided to crops and orchards by hoses, channels, sprays or drip systems in order to supplement rainfall

isobars lines on a map that join places with the same air pressure

literacy rate the proportion of the population aged over 15 who can read and write

liveable city a city that people want to live in, which is safe, well planned and prosperous and has a healthy environment

livestock animals raised for food or other products

location a point on the surface of the Earth where something is to be found

manufacture to makeproducts on a large scale

mental map a drawing or map that contains our memory of the layout and distribution of features in a place

meteorologist a person who studies and predicts weather

monsoon rainy season accompanied by south-westerly summer winds in the Indian subcontinent and South-East Asia

mosque place of worship for people who follow Islam (Muslims)

mound spring mound formation with water at its centre, which is formed by minerals and sediments brought up by water from artesian basins

natural disaster an extreme event that is the result of natural processes and causes serious material damage or loss of life

natural environment elements — such as wind, soil, flowing water, plants and animals — that influence the characteristics of an area

natural hazard an extreme event that is the result of natural processes and has the potential to cause serious material damage and loss of life

natural resources resources (such as landforms, minerals and vegetation) that are provided by nature rather than people

neighbourhood a region in which people live together in a community

non-government organisation non-profit group run by people (often volunteers) who have a common interest and perform a variety of humanitarian tasks at a local, national or international level

pastoral describes land used for keeping, or grazing, sheep or cattle

permafrost permanently frozen ground not far below the surface of the soil

place specific area of the Earth's surface that has been given meaning by people

polar vortex a large pocket of very cold air rotating in the same direction as the Earth's orbit

population density the number of people living in a square kilometre

precipitation rain, sleet, hail, snow and other forms of water that falls from the sky when water particles in clouds become too heavy

pull factors positive aspects of a place; reasons that attract people to come and live in a place

push factors reasons that encourage people to leave a place and go somewhere else

rainfall variability the change from year to year in the amount of rainfall in a given location

Ramadan the month of fasting in the Islamic calendar. It is a time for abstaining from food, drink and other physical needs during daylight hours, as well as reconnecting spiritually with God.

rebate a partial refund on something that has already been paid for

region any area of varying size that has one or more characteristics in common

relative humidity the amount of moisture in the air

remote describes a place that is distant from major population centres

ripple effect the flow-on effect of a particular action

riverine environment the environment around a river or river bank

run-off precipitation not absorbed by soil, and which runs over the land and into streams

sea change the act of leaving a fast-paced urban life for a more relaxing lifestyle in a small coastal town

soak place where groundwater moves up to the surface

southern oscillation a major air pressure shift between the Asian and east Pacific regions. Its most common extremes are El Niño events.

sparse thinly scattered or unevenly distributed; often used when referring to population density

storm shelter underground shelter where people can take refuge from a tornado

storm surge a sudden increase in sea level as a result of storm activity and strong winds. Low-lying land may be flooded.

subsistence farming a form of agriculture that provides food for the needs of only the farmer's family, leaving little or none to sell

temperate climate climate with generally warm summers and cool winters, without extremes

the Dreaming in Aboriginal spirituality, the time when the Earth took on its present form and cycles of life and nature began; also known as the Dreamtime. It explains creation and the nature of the world, the place that every person has in that world and the importance of ritual and tradition. Dreaming Stories pass on important knowledge, laws and beliefs.

Tornado Alley a region of the central United States across which tornadoes are most likely to form. The core states are Texas, Oklahoma, Kansas, Nebraska, eastern South Dakota, and the Colorado Eastern Plains.

torrential rain heavy rain often associated with storms, which can result in flash flooding

tree change the act of leaving a fast-paced urban life for a more relaxing lifestyle in a small country town, in the bush, or on the land as a farmer

troposphere the layer of the atmosphere closest to the Earth. It extends about 17 kilometres above the Earth's surface, but is thicker at the tropics and thinner at the poles, and is where weather occurs.

turbine a machine for producing power, in which a wheel or rotor is made to revolve by a fast-moving flow of water, steam or air

typhoon the name given to cyclones in the Asian region

undernourished not getting enough food for good health and growth

uranium radioactive metal used as a fuel in nuclear reactors

urban decay situation in which a city area has fallen into a state of disrepair through its people leaving the area or not having enough resources to look after it

virtual water all the water used to produce goods and services. Food production uses more water than any other production.

vulnerability the state of being without protection and open to harm

water footprint the total volume of fresh water that is used to produce the goods and services consumed by an individual or country

water scarcity a situation that occurs when the demand for water is greater than the supply available

water stress a situation that occurs in a country with less than 1000 cubic metres of renewable fresh water per person

water vapour water in its gaseous form, formed as a result of evaporation

weir a barrier across a river, similar to a dam, which causes water to pool behind it. Water is still able to flow over the top of the weir.

whiteout a weather condition where visibility and contrast is reduced by snow. Individuals become disoriented as they cannot distinguish the ground from the sky.

wilderness a natural place that has been almost untouched or unchanged by the actions of people

INDEX

Note: Figures and tables are indicated by italic *f* and *t*, respectively, following the page reference.